Andrew Lambert

PHYSICS

for First Examinations

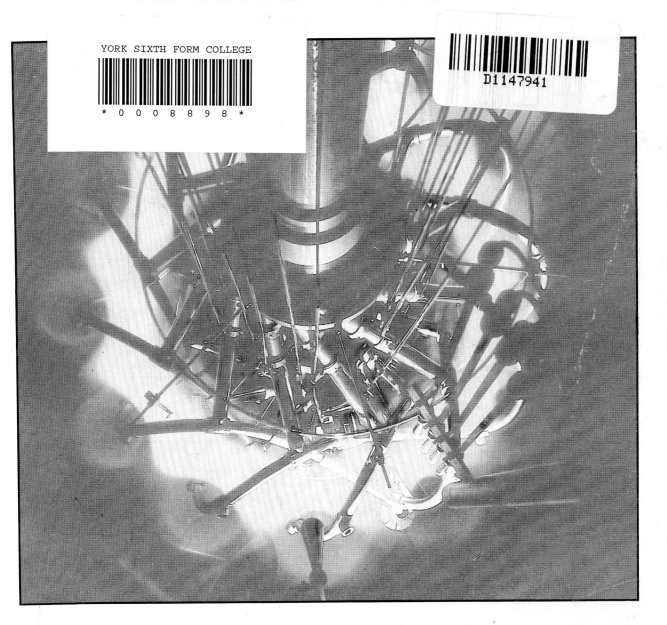

Blackie

Dedication
To Jean, Judith, Ruth and Nicholas

ISBN 0 216 91672 0
First published 1985

© Andrew Lambert 1985

Published by Blackie and Son Limited
Bishopbriggs, Glasgow G64 2NZ
Furnival House, 14–18 High Holborn
London WC1V 6BX

Cover photograph
A nuclear fusion experiment at Culham Laboratory,
Oxfordshire (see p. 255)

Filmset by Advanced Filmsetters (Glasgow) Ltd
Printed by Blantyre Printing and Binding Co. Ltd., Glasgow and London

PREFACE

Physics. What is it? What is it to do with you? Does it affect you? Is your life better off with or without it? What is it like to be a physicist? What do physicists do? How do they do it? Are physicists unusual people? Can you 'do' physics?

If you have read *Physics in Action* you might recognize these questions. They were in the first paragraph of the preface of that book. The rest of *Physics in Action* started answering those questions. You probably feel that you can begin to answer those questions. But there is a lot more to come, and this book takes over where *Physics in Action* stopped.

This book is full of questions. You should be able to write down the answers to some of the questions as soon as you read them. You will have to think hard and talk about some of the other questions with your friends or teacher.

At the end of each chapter there are a few more questions: 'Further Questions'. Some questions are to see whether you have understood what the chapter is about. Several questions have new ideas for you to think about.

Also at the end of each chapter is a summary of the main points covered in that chapter. These summaries are to help you when you are revising for an examination. You will find in these summaries some of the longer or more difficult words that have been used in the chapter. You must make sure you understand what each word means, or you will not be able to understand all of the chapter.

I hope you find this book useful, and enjoy using it.

Andrew Lambert
Ilkley

ACKNOWLEDGMENTS

The author and publishers would like to thank the following for permission to reproduce copyright material:

Fig. 1.1   British Petroleum Company p.l.c.

Fig. 1.9   Roger-Viollet

Fig. 1.10   Energy Technology Support Unit

Fig. 1.11   Wind Energy Group

Fig. 1.12   New Zealand High Commission

Fig. 4.1   Pictor International

Fig. 4.2   The J. Allan Cash Photolibrary

Fig. 4.3   NASA

Fig. 4.15, Fig. 5.3 and Fig. 9.18   All-Sport Photographic Ltd

Fig. 4.16   Austin-Rover Group Ltd

Fig. 5.8   Volvo Concessionaires Ltd

Fig. 5.10   Glasgow Herald and Evening Times Photolibrary

Fig. 6.3   Scottish Tourist Board

Fig. 6.20   Los Alamos National Laboratory and Science Photo Library

Fig. 8.1   Plumpton Agricultural College

Fig. 8.3   Central Office of Information

Fig. 8.10 and Fig. 8.11   Science Museum

Fig. 9.15   ICI p.l.c. Mond Division

Fig. 9.23   British Railways Board

Fig. 10.9   Imag

Fig. 10.12   Camera Press

Fig. 10.13   Frank Lane Picture Agency Ltd

Fig. 11.10 and Fig. 11.11   Historical Photography Collection, University of Washington Libraries, Photo by Farquharson

Fig. 11.12   Dundee Art Galleries and Museums

Fig. 11.15   Dr I. Grant, Heriot Watt University

Fig. 11.16   Hull City Council

Fig. 11.20   Boosey and Hawkes (Musical Instruments) Ltd

Fig. 12.29   Mr N. Jackson, Physics Department, Imperial College of Science and Technology

Fig. 13.29, Fig. 13.34 and Fig. 13.35   Royal Observatory, Edinburgh (© 1978, 1982 and 1984)

Fig. 13.43   British Telecom p.l.c.

Fig. 13.47   Lightdale/Science Source and Science Photo Library

Fig. 14.36   Yorkshire Electricity Board

Fig. 16.26   Fairchild Camera and Instrument (UK) Ltd

Fig. 17.1   CEGB (Midlands Region)

Fig. 17.34   Central Electricity Generating Board

Fig. 18.18   British Steel Corporation

Fig. 18.24   (c) British Alcan Lynemouth Ltd

Fig. 19.1   R. Wetmore and Science Photo Library

Fig. 19.2   R. K. Pilsbury

Fig. 19.17   Lawrence Berkeley Laboratory

Fig. 19.38   Ford of Europe Inc.

Fig. 19.39   CEGB Generation, Development and Construction Division

Fig. 20.1, Fig. 20.2, Fig. 20.29, Fig. 20.32, Fig. 21.40 and Fig. 21.41   United Kingdom Atomic Energy Authority

Fig. 20.24   The Royal Society

Fig. 20.25   C. T. R. Wilson and The Royal Society

Fig. 20.26   Radiation Laboratory, Berkeley, California

Fig. 20.27   Dr A. M. MacLeod, Dept. of Natural Philosophy, University of Glasgow

Fig. 20.28   (a), (b) and (c) P. M. S. Blackett and The Royal Society

Fig. 21.1   Jean Collombet and Science Photo Library

Fig. 21.36 and Fig. 21.37   Imperial War Museum

Fig. 21.42   UKAEA Culham Laboratory

Oxford and Cambridge Schools Examination Board:
page 181, Q8; page 239, Q7;
page 257, Q4; page 258, Q5

The cover photograph was supplied by UKAEA Culham Laboratory

Any photograph not acknowledged above is the copyright of the author.

CONTENTS

HAVE WE ENOUGH ENERGY ?

Fig. 1.1   BP's drilling rig Sedco in the North Sea

1.1   Introduction

North Sea oil! There are millions of barrels of it trapped in the rocks below the sea. Think of all the effort needed to find it and get it out from under the sea bed. Think of all the engineering problems involved. Oil must be important to deserve all that effort, mustn't it? It is a store of energy, and we need energy to make bricks to build our houses, to warm our houses, grow our food, cook our food, make our cars, move our cars—the list is endless. Most of the energy we need comes from **fossil fuels** like North Sea oil.

Why is oil called a 'fossil fuel'? Are there any other fossil fuels? How did fossil fuels get under the ground in the first place?

You need energy to keep alive. Where does *your* energy come from?

Our supply of fuels is rapidly dwindling. This will affect the way you live in the near future. One of the biggest challenges facing the human race is to find sources of energy that can take the place of fossil fuels. The solution to this problem will be a major contribution to our society, and physicists are one group of people in a position to help make that contribution.

In later chapters, you will learn what the laws of physics say you can and cannot do with energy, and learn how to *measure* some kinds of energy. This introductory chapter is about some of our *sources* of energy.

1.2 You should know . . .

Before you read this chapter, you should:

1 know about different *kinds* of energy, such as gravitational potential energy, chemical energy, internal energy, electrical energy;

2 know that heating is the name we give to the process of transferring internal energy from a hot body to a cooler body;

3 be able to describe simple models illustrating convection and conduction;

4 know that a hot body can give out radiant (or radiation) energy, such as infrared radiation or light;

5 know that when radiant energy is absorbed, the internal energy of the body is increased;

6 know what sorts of surface are good at giving out radiation, and what sorts of surface are good at absorbing it;

7 know that the important thing about energy is the way it can be changed from one kind to another—it is then that energy is useful to us.

Make a list of all the kinds of energy that you know about and give one example of each type. You might start your list with 'gravitational potential energy' and 'water stored in a high reservoir' as your example.

In Fig. 1.2(a), elastic potential energy is being changed to kinetic energy of the stone. Write down the energy changes which are taking place in the other photographs in Fig. 1.2.

(a)

(b)

(c)

(d)

(e)

(f)

Fig. 1.2   What energy changes do these photographs show?

1.3   What happens to fossil fuels?

Fig. 1.3   This power station burns fossil fuels

Fossil fuels are burned, to change the energy stored in them to other forms of energy. A fossil fuel may be burned in a **power station**, where the energy in the fuel is converted to electrical energy, or it may be burned in an **internal combustion engine** where the energy is converted to kinetic energy, as in a car.

Is all the energy made by burning a fuel in a power station or an internal combustion engine converted to useful energy? Obviously not; a lot goes up the chimney or out of the exhaust pipe, increasing the internal energy of the surrounding air. **Heat engines** *have* to 'throw away' some of the energy from the fuel to work at all. The reasons are complicated and are explained in the branch of physics known as **thermodynamics**.

Surely this 'waste' energy can be used again? Unfortunately, it is not simple to do so, and certainly does not provide an answer to our energy problems. This energy is spread out thinly over many billions of molecules in the atmosphere, each molecule having only a small share of the energy. To 'concentrate' this energy again so that it can be useful needs a machine which, of course, needs a fuel! Such machines do exist; they are sometimes called **heat pumps**. They provide a comparatively efficient way of using a fuel for heating buildings.

We are burning our store of fossil fuels at an ever increasing rate, and nearly all the energy becomes 'diluted' energy that we cannot use. It is a one-way process. How long will our fossil fuels last? Nobody is sure, but most estimates suggest that supplies of oil will be running very short by the year 2010. (How old will you be then?) Coal will last at least another 200 years.

Not only do we use coal and oil as a fuel, but they are also the raw materials for a wide range of products of the chemical industry, such as drugs, medicines, perfumes, plastics and man-made fibres.

Some of the remaining sections in this chapter suggest other sources of energy which might eventually replace fossil fuels.

1.4   Energy from the Sun

Every second, the Earth receives from the Sun 10^{17} joules of energy. (The number 10^{17} is 1 followed by 17 noughts, or one hundred thousand million billion! A 'joule' is a unit of energy. You will learn later exactly how much energy is in 1 joule. It is roughly sufficient to increase the temperature of $\frac{1}{4}$ g of water by 1 °C.)

Also, in every second the world uses an average of 10^{14} joules of energy, most of which comes from burning fossil fuels. The Earth is receiving from the Sun 1000 times as much energy as we actually use—and it is free! Surely we could make use of some of this energy from the Sun rather than burn so much fossil fuel?

Fig. 1.4   The Earth's main source of energy—the Sun

One difficulty in making direct use of the energy from the Sun is that the 10^{17} joules which arrive every second are spread out over half the Earth's surface at any one time, the average being a few hundred joules per second on every square metre, and that includes large areas of ocean and desert. However, increasing use is being made of energy from the Sun to heat homes and provide hot water.

Fig. 1.5   Solar panels

Fig. 1.5 shows **solar panels** fixed to the roof of a house to 'capture' the Sun's energy. Let us see how these solar panels work.

Fig. 1.6 is a diagram of a typical system which uses a solar panel to provide hot water.

Energy from the Sun travels to the Earth in the form of **electromagnetic radiation**. This is an important kind of wave. (You will learn more about electromagnetic radiation in Chapter 12.) You should know that a lot of this radiation is **light**, or **visible radiation** (You can see it!), and that this visible radiation is made up of all the colours of the rainbow. You should also know that a lot more of this radiation is invisible radiation called **infrared radiation**. If you *could* see infrared radiation, where-abouts in the rainbow would it be?

Much of this radiation passes through the glass at the front of the solar panel and hits the surface at the back. This surface is *black*; black surfaces are good at *absorbing* radiation (both visible and invisible). The internal energy of the black surface increases and it becomes hot.

What do white or silvery surfaces do to radiation? Why do people wear white clothes in hot countries?

You should know that black surfaces are also good at *giving out* radiation if they become hot. The black surface of the solar panel *is* becoming hot, so it 're-radiates' some of the energy, but this radiation is of much longer wavelength than that coming from the Sun. Some of it is absorbed by the glass, and some is reflected back. What happens to the glass?

Energy is *conducted* along the back of the solar panel to the copper pipes. Why are the pipes made of *copper*? What should the back of the solar panel be made of?

The pipes are filled with a liquid which becomes hot. Oil is often used as the liquid. Why do you think oil is a better liquid to use than water?

Hot liquids rise because they expand as they warm up and so they become less dense. The hot oil rises up by *convection* to the **energy exchanger** in the hot water tank. Energy is conducted through the pipes to the cold water in the tank. Usually a pump is used to help the natural convection of the oil. The pump only switches on if the oil is hotter than the water in the storage tank.

This system is often used to pre-heat domestic hot water, which is then heated further by burning a fossil fuel, such as gas. However, on a warm day in Britain the solar system alone can produce a water temperature of 60 °C.

Why is the energy exchanger at the bottom of the cold water tank?

Why is insulation often wrapped round the pipes? Suggest a suitable substance for this insulation.

What disadvantages does solar heating have?

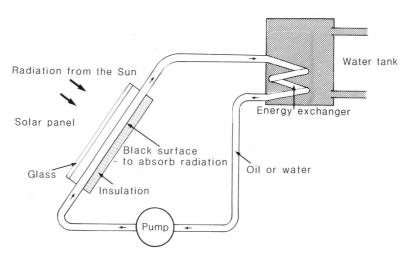

Fig. 1.6   A solar heating system

1.5 Energy from water

The probability that we will run out of fossil fuels has led to a lot of research into other natural sources of energy. You looked at the way the Sun's energy can be used for direct heating in the last section. There are several other possibilities.

For example, there is **hydroelectric power**. This source of energy has already been used for many years. Water in a reservoir converts its gravitational potential energy to electrical energy as it falls to the generating station. Fig. 1.7 shows the turbine hall of a small hydroelectric power station in Scotland. The turbine furthest from the camera was being overhauled when the photograph was taken.

Fig. 1.7   The turbine hall of a small hydroelectric power station

How did the water get to the reservoir in the first place? Where did it get its energy from?

About 1% of the world's energy supply comes from hydroelectric power. Most of the suitable sites in Britain have already been used. A disadvantage of hydroelectric power is that suitable sites are often a long way from where the electricity is needed. Why do you think this is? (Suitable sites are usually in mountainous country. Where do most people live?)

Pumped storage systems are a development of hydroelectric power generation. The Ffestiniog power station illustrated in Fig. 1.8 is an example of this kind of scheme. The big coal-fired power stations in, for example, Yorkshire and Nottinghamshire, produce electricity fairly cheaply. It is very difficult to shut down these power stations and start them up again quickly, so during the night they produce more electricity than is actually needed. Some of this electricity is used by Ffestiniog power station to pump water from the Tan-y-Grisiau reservoir at the bottom of the picture to the storage reservoir which is behind the dam wall at the top

Fig. 1.8   Ffestiniog pump storage power station

left of the picture. During any period of the day when there is a heavy demand for electricity, water flows from the top reservoir through the power station and generates electricity in the usual way. The cheap electrical energy produced by the large, efficient modern power stations is stored (in the form of gravitational potential energy) in the top reservoir until it is needed.

Tidal power is a form of hydroelectric power. The rise and fall of the tides can fill and empty large reservoirs. The flow of water into and out of these reservoirs can drive turbines as in an ordinary hydroelectric power station.

What causes the tides?

Fig. 1.9   The tidal power station at La Rance in France

The only tidal power station in existence so far (1985) is at La Rance in France. This is designed to generate an average of 65 megawatts (MW) of power, but the output fluctuates. (Why do you think this is?) This power station has given a lot of problems. Try to find out whether it is still working.

The only likely site for tidal power in Britain is in the Severn estuary, where the rise and fall of the tides is sufficient for such a scheme to be feasible. The CEGB estimates that such a scheme could provide over 10% of the country's present electricity needs, but it would be very expensive to install and would be bound to affect the ecology of the Severn estuary.

1.6 Other sources of energy

There are a number of other suggestions that have been made for making use of natural energy sources. Here are brief details of some of them.

Wave energy

A lot of energy is stored in the waves on the sea, and devices have been developed that can extract up to 90% of this energy. It has been estimated that a line of these devices 1000 km long, moored off the west coast of Britain, could provide up to $\frac{1}{5}$ of the energy we need. Imagine the cost of 1000 km of these devices!

Fig. 1.10   One design for a device to extract energy from the waves

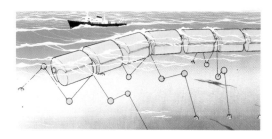

Fig. 1.12
A geothermal power station at Wairakei, New Zealand

Wind energy

In Britain there used to be over 10 000 windmills similar to the one in Fig. 1.2. What were they used for? They did not generate much power—perhaps 200 watts (W). Large windmills have been built in recent years; Fig. 1.11 shows a windmill on Orkney designed to produce up to 300 kilowatts (kW) of electricity. This windmill was built during 1983. To use wind power on a large scale would create problems. A lot of windmills, well spaced out, would be needed. They would certainly spoil the scenery and they would take up valuable space. Also, it is not always windy!

Geothermal energy

Many of the volcanic regions of the Earth are known for their hot springs. The energy to heat the water comes from the hot rocks under the surface. The hot water and steam can be used to provide home heating or to drive turbines in power stations. Iceland, Italy and New Zealand are three countries which use geothermal energy.

Nuclear energy

This is becoming an increasingly important source of energy. You will learn more about nuclear energy in Chapter 21.

Fig. 1.11   A windmill on Orkney for generating electricity

1.7 Where are you going?

It must be clear to you that physicists and engineers need to know about energy—where it comes from, how to use it efficiently and how to measure it, before they can begin to think of sensible ways of providing us with the energy we will demand in the future. As you read this book, you will learn, for example, how to *measure* energy and how *machines* convert energy from one form to another.

From time to time you may read newspaper or magazine articles, or see television programmes, on the subject of energy. You may read that hundreds of windmills will solve our energy problems, or that what we need are many nuclear power stations. By the time you have finished this book you should be able to understand the arguments in these articles and programmes. You should be able to decide for yourself whether the arguments are sensible ones. You should also learn a lot of other physics as well!

SUMMARY

Now that you have finished studying this chapter on sources of energy, there are a number of things you should know and be able to do.

1 You should:
  a) know that our supply of fossil fuels will not last very much longer;
  b) know some of the possible alternative sources of energy, and have an idea of what contribution they can make to our energy supply;
  c) know how a solar heating system works;
  d) know that black surfaces both absorb and emit radiation energy well, while white surfaces reflect radiation energy well.

2 You should know what each of the following is, or what each does:

fossil fuel	power station
internal combustion engine	solar panel
infrared radiation	visible radiation
reservoir	energy exchanger
	geothermal energy

3 These are some of the other words that have been used in this chapter. You should know what each word means:

source	dwindle
challenge	alternative
contribution	convert
complicated	ultimately
concentrate	insulation
ecology	fluctuate
install	feasible

FURTHER QUESTIONS

1 A large proportion of fossil fuels is burned in order to heat homes, factories, offices and schools. In what ways could your school save fuel without anyone suffering from the cold?

2 Make a list of some of the ways in which you see energy being wasted every day. Can you think of any ways in which this wasted energy could be saved?

3 a) What kind of energy from the Sun falls on a sunbather?
  b) What happens to this energy when it reaches the sunbather?
  c) Energy continues to fall on the sunbather all the time, but the temperature of the sunbather's body does not continue to increase. Explain why.

4 Explain all the energy changes involved as energy from the Sun becomes energy given out by an electric fire which is supplied with electricity from a hydroelectric power station.

5 Think of as many different sources of energy as you can, which do not involve burning fossil fuels. For each source of energy, make a list of as many of its advantages and disadvantages that you can think of.

2

MEASURING MOVEMENT

Fig. 2.1

Fig. 2.4

Fig. 2.2

Fig. 2.5

Fig. 2.3

Figs 2.1–2.8   Some of the many ways in which things can move

2.1   Introduction

Look at the photographs on these pages—they are all photographs of something *moving*. Some things are moving fast, some more slowly. Some things keep moving in the same direction for a while—can you see anything which does this? Can you see anything which continually changes its direction? Is there anything which moves backwards and forwards?

Look around you. You can probably see something moving. All these moving things have energy —**kinetic energy**. An important part of physics is about things which move, describing how they move and why they move, and measuring the movement.

Other people apart from physicists need to know about movement. A team of engineers at Derby designed the high-speed train in Fig. 2.1. Not only is the high-speed train as a whole moving in a straight line, but the engines which move it have many moving parts themselves. Some of the parts are moving round, others are moving backwards and forwards. Engineers need to know, among many other things, the rules and laws of movement (or **dynamics**, to give this area of physics its proper title) in order to be able to do their job properly.

Fig. 2.6

Fig. 2.7

Fig. 2.8

You need to be able to describe *how* things move, and know how to *measure* the movement, before you can explain *why* things move. You also need to know about movement before you can learn about kinetic energy. This chapter is about measuring the movement of things which are travelling in the same direction all the time. The high-speed train was travelling in a straight line when the photograph was taken, although it will not keep doing so for ever! You will have a closer look at round-abouts (and other things going round in circles) and swings (and other things which go backwards and forwards) in a later chapter.

This chapter is also about **graphs**. Scientists make a lot of use of graphs to show experimental results, for example, in a way that others can easily understand. You must be able to draw graphs quickly and accurately.

2.2 You should know . . .

To understand this chapter, you should:
1 know how to measure speed using a stopwatch and ruler;
2 know the units with which we measure speed;
3 be able to use a ticker-tape timer and make a ticker-tape chart;
4 know how to measure speed and acceleration from a ticker-tape chart;
5 know the meaning of the phrases 'constant speed', 'average speed', 'constant acceleration' and 'average acceleration'.

In case you have forgotten, or are not sure, of these points, here is a quick reminder of some of them.

Speed

The speed of anything is the distance it travels in one unit of time—1 second (s), 1 minute (min), etc. The **units** we usually use in the lab are **metres per second**. (This is abbreviated to 'm/s', but you may

also see it written as 'ms^{-1}'.) Speeds of cars and trains are often measured in kilometres per hour (km/h).

To measure the speed of the hang glider in Fig. 2.4 you need to measure how far it travels and the time it takes to travel that distance. At the time the photograph was taken, the hang glider moved 75 m in 25 s. It therefore went $75/25 = 3$ metres in each one of those seconds. Its speed was 3 m/s.

Generally, you can calculate the **average speed** of anything from

$$\text{average speed} = \frac{\text{total distance travelled}}{\text{time taken}}$$

Acceleration

Most of the things illustrated so far in this chapter are likely to *change* their speed rather than move at a constant speed. Look at the boys running a 100 m sprint in Fig. 2.2. At the moment the photograph was taken the leader was running at about 8 m/s. He was obviously not running at all just before the race. Suppose he took 1 second to increase his speed from nothing (when the starting pistol went) to 8 m/s. We say his **acceleration** was 8 m/s in 1 second. We would normally write this as 8 m/s^2 (or 8 ms^{-2}). Learn how to write the units of acceleration correctly.

Suppose the hydrofoil boat in Fig. 2.3 was moving at 10 m/s when the photograph was taken, and that 5 seconds later it was moving at 25 m/s. In that 5 seconds it would increase its speed by 15 m/s $(25-10 = 15)$. It must have increased its speed by an average of 3 m/s in each second $(15/5 = 3)$. The boat's average acceleration was 3 m/s^2. Why do you think this is an *average* acceleration?

Generally, you can calculate acceleration from:

$$\text{acceleration} = \frac{\text{change of speed}}{\text{time taken for speed change}}$$

Ticker-tape charts

Fig. 2.9   A ticker-tape timer

The ticker-tape timer (Fig. 2.9) hammers dots onto a piece of tape, usually at the rate of 50 per second, thus leaving a permanent record of the way something moves. The tape can be cut up, for example, every 10 spaces (Fig. 2.10); each length will show how far the object moved in $\frac{1}{5}$ second. The lengths can be stuck side by side to make a chart as in the photograph.

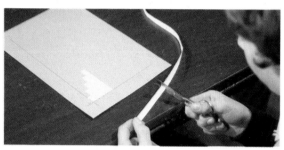

Fig. 2.10   Making a tape chart

Fig. 2.11 reminds you how to measure the average velocity shown by one piece of tape, and also how to measure the acceleration shown by a tape chart.

Why use a ticker-tape timer at all? What is wrong with a stopwatch? Do you think a ticker-tape timer has any advantages over a stopwatch? In later chapters you will learn that there are many other ways of obtaining *data* about moving objects apart from using a ticker-tape timer.

2.3   Speed and velocity

As you know, the speed of something is the distance it travels in one unit of time (e.g. 1 second). You may also have heard of the word 'velocity'; does this mean the same thing? Not quite—there is an important difference between the two words.

Fig. 2.12

The canoe in Fig. 2.12 is moving south at 3 m/s. In the next photograph it has been joined by a second canoe, also moving south at 3 m/s, so the two canoes have the same speed. Because they are

Fig. 2.11   How to calculate the acceleration from a tape chart

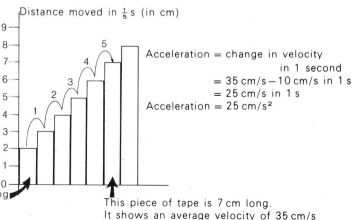

Distance moved in $\frac{1}{5}$ s (in cm)

Acceleration = change in velocity in 1 second
= 35 cm/s − 10 cm/s in 1 s
= 25 cm/s in 1 s
Acceleration = 25 cm/s²

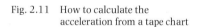

This piece of tape is 2 cm long.
The tape moved 2 cm in $\frac{1}{5}$ s.
It would go 10 cm in 1 s.
The average velocity shown by this piece of tape is 10 cm/s

This piece of tape is 7 cm long.
It shows an average velocity of 35 cm/s

Fig. 2.13

moving in the same *direction*, we *also* say that they have the same **velocity**. If the second canoe then turns east but continues to move at 3 m/s, the two canoes still have the same speed, but they now have *different* velocities because they are moving in different directions.

For two things to have the same velocity, not only must they travel the same distance in each unit of time, but they must also travel in the same direction:

$$\text{velocity} = \frac{\text{distance moved in a particular direction}}{\text{time taken}}$$

Suppose the high-speed train in Fig. 2.1 goes along a curved section of track at a steady speed of 10 m/s. Its *speed* is the same all the time, but its *velocity* is changing all the time, because the direction in which it is moving is changing all the time.

Quantities like velocity which must have a direction specified are called **vector quantities**. Quantities which do not have a specified direction are called **scalar quantities**. Speed is a scalar quantity. Temperature is another scalar quantity. Write down any other examples of scalar quantities that you can think of.

2.4   Velocity-time graphs

Although a chart like the one Michael was making in Fig. 2.10 tells you a little about the way something was moving, scientists generally find a **line graph** much more useful. You will see later in this chapter that you can find out more about the way something is moving from a line graph than from a tape chart.

A chart such as the one in Fig. 2.11 can be converted to a line graph quite simply by joining the tops of each length of tape, as shown in Fig. 2.14.

Notice the following points about this graph.
1 Both axes are labelled with:
   a) what is being plotted;
   b) the units that are being used.
2 Both axes have numbers, evenly spaced.
Make sure that all the graphs you draw have these features.

The length of each piece of tape is the distance travelled in $\frac{1}{5}$ second, so the average velocity represented by each piece of tape is five times the length of the tape. Every centimetre on the *velocity axis* represents a velocity of 5 cm/s.

Why has the *time axis* been marked out in $\frac{1}{5}$ second intervals on the diagram?

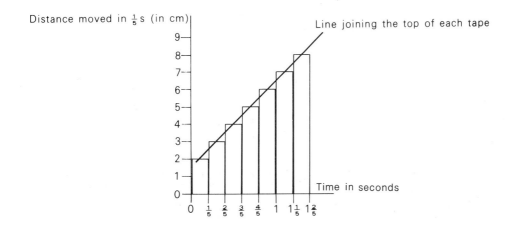

Fig. 2.14   Converting a tape chart to a line graph

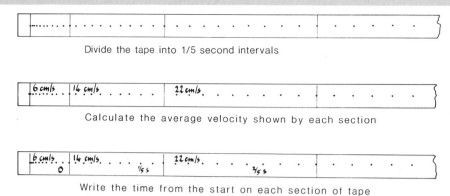

Divide the tape into 1/5 second intervals

Calculate the average velocity shown by each section

Write the time from the start on each section of tape

Time in seconds	0	⅕	²⁄₅	³⁄₅	⁴⁄₅	1	
Velocity in cm/s	6	14	22				

Transfer the data to a table, from which you can draw a graph

Fig. 2.15   Drawing a line graph from a ticker tape record

You do not *have* to make a tape chart before you draw a line graph. You need only mark out the tape in, say, $\frac{1}{5}$ second intervals, calculate the velocity shown by each length of tape and draw a graph from these figures. Fig. 2.15 explains this in more detail.

Fig. 2.16   How do you measure this trolley's acceleration?

2.5   Using a velocity-time graph to find acceleration

A trolley running down a slope, like the trolley in Fig. 2.16, accelerates. How can you measure the acceleration? It is possible to buy an **accelerometer** which is an electronic device for measuring acceleration directly. It is very unlikely that the trolleys you use are fitted with one!

You can easily use a velocity-time graph to find the *acceleration* of a trolley. Assemble the apparatus illustrated in the photograph. Pass a length of tape through the ticker-tape timer, stick it to the trolley,

start the timer, and release the trolley from the top of the slope. Use the tape to make a velocity-time graph in one of the ways described in the last section.

To use your graph to find the acceleration of your trolley, you must remember that acceleration is the change of velocity in 1 second. Read the velocity shown by your graph at or near the start. Then read the velocity 1 second later. The difference between these two velocities is the acceleration. Fig. 2.17 shows you how.

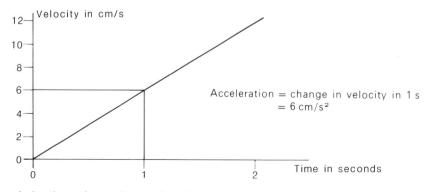

Acceleration = change in velocity in 1 s
= 6 cm/s²

Fig. 2.17   How to calculate the acceleration from a velocity-time graph

What was the acceleration of your trolley?

If your graph is a straight line graph like Fig. 2.17 then the acceleration of your trolley stays constant. If your graph was not a straight line then you will have calculated the *average* acceleration.

You do not *have* to use a time of 1 second when calculating the acceleration from a graph. Remember that

$$\text{acceleration} = \frac{\text{change in velocity}}{\text{time taken for velocity to change}}$$

Fig. 2.18 shows how you can use a velocity-time graph to calculate the acceleration over a period of 3 seconds.

A *steep* velocity-time graph shows a *large* acceleration (Fig. 2.19) while a *less steep* graph shows a *smaller* acceleration (Fig. 2.20).

The 'steepness' of a slope can be measured, as the photograph of the road sign in Fig. 2.21 should remind you. A slope of '1 in 6' means that for every 6 metres you travel *along*, you go *up* 1 metre. The steepness of a graph could be measured in the same way.

We do not usually talk about the 'steepness' of a graph; we refer to either the **slope** or the **gradient** of a graph. You calculate the gradient of a graph from

$$\text{gradient} = \frac{\text{increase in the } y\text{-direction}}{\text{corresponding increase in the } x\text{-direction}}$$

What is the gradient of the velocity-time graph in Fig. 2.17? You have just seen that the gradient is the increase in the *y*-direction (in this case, the increase in the velocity) for every increase of 1 unit in the *x*-direction (in this case, for every 1 second). But this is *exactly* how you calculated the acceleration from the velocity-time graph. So a very useful fact about velocity-time graphs is:

The gradient of a velocity-time graph is equal to the acceleration.

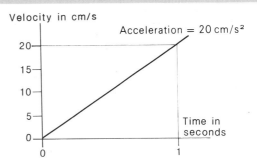

Fig. 2.19

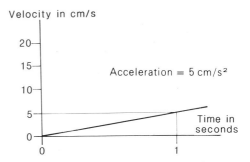

Fig. 2.20

Fig. 2.21

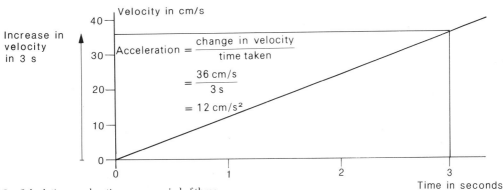

Fig. 2.18   Calculating acceleration over a period of three seconds

2.6 Using a velocity-time graph to find the distance travelled

How far did your trolley move in, say, the first 3 seconds? You can obtain this information from your velocity-time graph as well. Remember that you can calculate how far anything travels from

distance travelled = average speed × time taken

What is the average speed shown by the graph of your trolley's motion? Look at Fig. 2.22. (This is the same graph as was drawn in Fig. 2.18.) At the start the velocity was 0 cm/s. After 3 seconds the velocity was 36 cm/s. The average velocity is

$$\frac{0+36}{2} = 18\,\text{cm/s}$$

In other words, the average velocity is half the 'final' velocity (the velocity after 3 seconds). Therefore:

$$\begin{aligned}
\text{distance travelled} &= \text{average velocity} \times \text{time} \\
&= \tfrac{1}{2} \times \text{final velocity} \times \text{time} \\
&= \tfrac{1}{2} \times 36\,\text{cm/s} \times 3\,\text{s} \\
&= 54\,\text{m}
\end{aligned}$$

Use this method with *your* graph to calculate the distance travelled by your trolley in 3 seconds. If your graph is not a straight line for 3 seconds, use a time of 1 or 2 seconds instead.

On a graph like the one in Fig. 2.22 you can draw a line upwards from the 3 second mark until it meets the line of the graph, as has been done in Fig. 2.23. There is then a *triangle* on the graph paper, shown shaded in Fig. 2.23. The *area* of this triangle is calculated from

$$\text{area of triangle} = \tfrac{1}{2} \times \text{base} \times \text{height}$$

This triangle has a height of 36 units and a base of 3 units, so

$$\begin{aligned}
\text{area of triangle} &= \tfrac{1}{2} \times 36 \times 3 \\
&= 54
\end{aligned}$$

But this is exactly how we calculated the distance travelled. Therefore:

The 'area under the velocity-time graph' is equal to the distance travelled.

Although you have only seen this to be true for this one case, it is in fact true for all velocity-time graphs, no matter what their shape—they do not have to be straight lines.

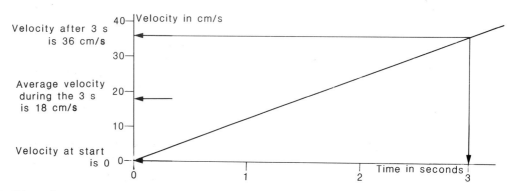

Fig. 2.22 What is the average speed shown by this graph?

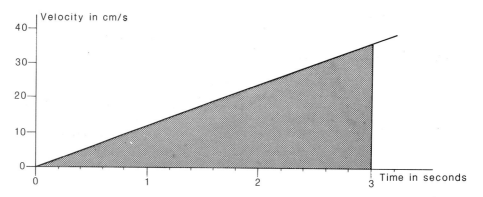

Fig. 2.23 Measuring the area under a graph

The idea of the 'area under a graph' is an important one to physicists. You will find that it is useful with many other graphs as well as velocity-time graphs.

Make sure you understand what we mean by the 'area under a graph', and know how to find it.

2.7 Distance-time graphs

Another graph you might need to draw when dealing with moving objects is a **distance-time graph**. How do you think you would use a ticker-tape record to draw a graph of distance travelled (from the start of the runway, say) against time?

You learned in the last two sections how to obtain useful information from a velocity-time graph. What useful information? Can you obtain any useful information from a distance-time graph?

It turns out that the area under a distance-time graph has no meaning, but the *gradient* of the graph gives you some useful information.

Look at Fig. 2.24, which shows a distance-time graph. The gradient is the increase in the 'y' reading (distance in this case) divided by the corresponding increase in the 'x' reading (time in this case). But distance divided by time is exactly how you calculate the velocity of a moving object. Therefore:

The gradient of a distance-time graph is equal to the velocity.

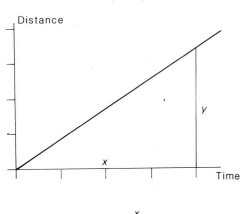

$$\text{gradient} = \frac{x}{y}$$

$$= \frac{\text{distance}}{\text{time}}$$

$$= \text{velocity}$$

Fig. 2.24   A distance-time graph

By now you should be able to draw and recognize the shapes of velocity-time graphs and distance-time graphs for various different sorts of movement. Draw the shape of both the distance-time graph and the velocity-time graph for something that is moving with:
1) constant velocity;
2) constant acceleration.

2.8   An equation of motion

Drawing and using a graph are important skills for a physicist, but drawing a graph takes some time. It is not always necessary to draw a graph, as you will learn in the rest of this chapter.

Remember that you can calculate acceleration from

$$\text{acceleration} = \frac{\text{change in velocity}}{\text{time taken}}$$

Scientists call this an **equation**, in this case *relating* acceleration, change in velocity and time. Rather than writing out this equation using words, we usually use a single letter to represent each of the quantities. This equation could be written

$$a = \frac{v-u}{t}$$

where a stands for acceleration
t stands for time taken
v stands for velocity at the end of this time (often called the 'final velocity')
u stands for velocity at the start of this time (often called the 'initial velocity').

So $v-u$ is the *difference* between the final and initial velocities, in other words, the *change in velocity*.

You will often see this equation written in a different way. We can use a little algebra, as follows. Multiply both sides of the equation by t:

$$at = v-u$$

and rearrange the equation:

$$v = u+at$$

All this is saying is that you can calculate the 'final' velocity v by multiplying the acceleration a by the time taken t, and adding it to the 'initial' velocity u.

This is known as one of the **equations of motion**. Remember that it only applies to bodies moving with *constant acceleration*.

Here is an example of how to use this equation. Fig. 2.25 shows Ruth just starting to come down a slide. Suppose she starts from rest at the top and accelerates at 1 m/s². If she takes 3 seconds to come down the slide, how fast will she be moving at the bottom?

Fig. 2.25 How fast will Ruth be moving at the bottom of this slide?

One of the secrets of successfully managing calculations in physics is to lay your answer out neatly and to write down exactly what you are doing. It helps you to understand, and it enables other people to follow what you are doing.

First, write out a *table* of the information that you have; do not forget to include the *units* with any numbers.

initial velocity $u = 0$ (Ruth starts from rest.)

acceleration $a = 1$ m/s²
time $t = 3$ s
final velocity $v = ?$ (This is what we are being asked to find.)

Next, decide which equation to use to solve the problem. (Sometimes you need more than one.) Since we only know one equation so far, this is an easy decision! Write down the equation you are going to use:

$$v = u + at$$

Now, using the table of information, put the correct numbers and units into the equation. Notice that the units should be written as well as the numbers:

$$v = 0 + 1 \text{ m/s}^2 \times 3 \text{ s}$$

Now do the calculation:

$$v = 3 \text{ m/s}$$

Finally, write out the answer in the form of a sentence:

'Ruth will be travelling at a speed of 3 m/s when she reaches the bottom of the slide.'

It is very important that you set out answers to problems like this in a way that other people can understand when they read your paper. The most brilliant scientists are useless if they cannot explain to other people what they are doing.

2.9 Another equation of motion

How far does Ruth travel in the 3 seconds for which she was accelerating down the slide? (In other words, how long is the slide?) The next equation of motion enables you to calculate this distance, without having to find the area under a velocity-time graph.

Remember that you can calculate the distance travelled from

distance travelled = average velocity × time taken

or, in letters,

$$s = v_{av} \times t$$

(We use the letter s to represent the 'distance travelled'.)

The average velocity, v_{av}, is the average of the initial and final velocities, so

$$v_{av} = \tfrac{1}{2}(v + u)$$

We can rewrite the equation above as

$$s = \tfrac{1}{2}(v + u) \times t$$

But you know, from the first equation of motion, that

$$v = u + at$$

so where there is v in our equation for the distance travelled, we can write $u + at$ instead, like this:

$$s = \tfrac{1}{2}[(u + at) + u] \times t$$

Adding together the two u's, and multiplying out the brackets, gives

$$s = ut + \tfrac{1}{2}at^2$$

This is the second equation of motion.

Now we can answer the question about how long the slide is. Setting out the answer in the same way as in the previous example:

$$\begin{aligned}
\text{initial velocity} \quad & u = 0 \\
\text{acceleration} \quad & a = 1 \text{ m/s}^2 \\
\text{time} \quad & t = 3 \text{ s} \\
\text{distance} \quad & s = ?
\end{aligned}$$

Using
$$\begin{aligned}
s &= ut + \tfrac{1}{2}at^2 \\
s &= 0 + \tfrac{1}{2} \times 1 \text{ m/s}^2 \times 3 \text{ s} \times 3 \text{ s} \\
&= 4.5 \text{ m}
\end{aligned}$$

The slide is 4.5 m long.

2.10 A third equation of motion

In order to use both the equations you have seen so far you need to know the time t. Sometimes it is convenient to have an equation that you can use without needing to know the time taken. The third equation of motion does not involve time. It can be derived from the others by using a little algebra. Do not worry about learning the algebra which follows, although if you can understand it so much the better. It is the result that is important.

The first equation of motion you learned about was

$$v = u + at$$

We can rearrange this so that t is the 'subject' of the equation:

$$t = \frac{v - u}{a}$$

Now, in the second equation, everywhere there is a letter t we can write $\dfrac{v - u}{a}$ instead:

$$s = \frac{u(v - u)}{a} + \tfrac{1}{2}a\left(\frac{v - u}{a}\right)^2$$

Cancelling one of the a's in the last part of the equation, and then multiplying all through by $2a$, gives

$$2as = 2u(v - u) + (v - u)^2$$

Multiplying out the brackets gives

$$2as = 2uv - 2u^2 + v^2 - 2uv + u^2$$

Collecting up the like terms and rearranging the equation gives

$$v^2 = u^2 + 2as$$

Here is an example of this equation of motion being used.

The wall off which Toby is jumping in Fig. 2.26 is 1.5 m high. The acceleration downwards (due to gravity) is 10 m/s². With what velocity does Toby hit the ground?

Fig. 2.26   With what velocity does Toby hit the ground?

Setting out the calculation in the same way as before

$$\begin{aligned}
\text{initial velocity} \quad & u = 0 \\
\text{distance} \quad & s = 1.5 \text{ m} \\
\text{acceleration} \quad & a = 10 \text{ m/s}^2 \\
\text{final velocity} \quad & v = ?
\end{aligned}$$

Using
$$\begin{aligned}
v^2 &= u^2 + 2as \\
v^2 &= 0 + 2 \times 10 \text{ m/s}^2 \times 1.5 \text{ m} \\
&= 30 \text{ (m/s)}^2 \\
v &= 5.5 \text{ m/s}
\end{aligned}$$

Toby will hit the ground at a velocity of 5.5 m/s.

Notice that the square root of 30 is actually 5.4772256 (to 8 significant figures) but that the answer has been rounded to 2 significant figures, because the data has been given to 2 significant figures. Always round off your answers in this way. Two significant figures is usually sufficient; no problem in this book will require more than three.

SUMMARY

Now that you have finished studying this chapter on motion, there are a number of things you should know and be able to do.

1 You should:
a) know that

$$\text{speed} = \frac{\text{distance travelled}}{\text{time taken}}$$

b) know the difference between speed and velocity;

c) know that

$$\text{acceleration} = \frac{\text{change in velocity}}{\text{time taken}}$$

d) know how to draw a velocity-time graph and a distance-time graph;

e) know what is meant by the gradient of a graph, and how to measure the gradient;

f) know what is meant by the area under a graph, and how to measure the area;

g) know that the gradient of a distance-time graph is equal to the velocity;

h) know that the gradient of a velocity-time graph is equal to the acceleration;

i) know that the area under a velocity-time graph is equal to the distance travelled.

j) know how to write the units of velocity and acceleration correctly;

k) know how to use the three equations of motion;

l) know that you must round off your answers to numerical questions to a suitable number of significant figures.

m) be able to set out a calculation so that somebody else understands what you are doing.

2 You should know what each of the following is, or what it does:

ticker-tape timer	tape chart
scalar	dynamics
data	axes (of graph)
equation	square root
line graph	

3 These are some of the other words that have been used in this chapter. You should know what each word means:

constant	average
quantity	plot
regular	interval
increase	specified
transfer	calculate
corresponding	initial
represent	derive
sufficient	obtain

FURTHER QUESTIONS

1 In Section 2.1 at the start of this chapter you read that the engineers who designed the high-speed train needed to know about the laws of motion. Suggest two or three other areas of physics that these engineers would need to know about.

2 The table below shows how fast a car was moving at various times after the start of its journey.
a) From these figures, draw a graph of velocity against time.
b) Between what two times was the acceleration constant?
c) What was the value of the constant acceleration? Explain how to obtain your answer.

Velocity of car (m/s)	0	1.0	2.0	3.5	5.0	7.0	10.0	13.0	16.0	19.0	21.0	23.0	24.0
Time (s)	0	2	4	6	8	10	12	14	16	18	20	22	24

3 The graph in Fig. 2.27 is a speed-time graph of a boy running to school.
 a) By looking at the graph, describe his journey to school.
 b) Use the graph to calculate the distance from the boy's home to school.

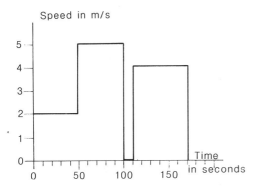

Fig. 2.27

The graph in Fig. 2.28 is the speed-time graph of a girl cycling home from school.
 c) From the graph, describe her journey.
 d) Use the graph to calculate her acceleration as she first left school.
 e) Use the graph to find the distance from school to her home.

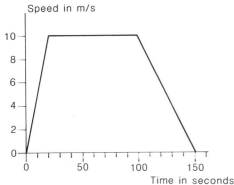

Fig. 2.28

4 Fig. 2.29 shows the speed-time graph of a car taking a commuter to work.
 a) By looking at the graph, do you think it likely that the commuter lives and works in the same town, or in different towns a few kilometres apart? Give reasons for your answer.
 b) Use the graph to find the acceleration during each of the following parts of the journey:
  i) A to B
  ii) B to C
  iii) C to D
  iv) D to E
 c) What was the distance covered by the car when travelling from C to E?
 d) A real car, travelling along a real road, would not have a simple speed-time graph like the one illustrated. Why not?

5 Draw a speed-time graph of your journey home from school. You will have to make sensible estimates of the speed at which you walk, or the speed of the bus or train. A brisk walking speed is 2 m/s. Thirty miles per hour is roughly equal to 15 m/s. Label some of the features of your graph, e.g. waiting at a bus stop, running for a bus, gazing in a shop window.

6 A train starts from rest at a station and accelerates with a constant acceleration of 0.5 m/s². Use suitable equations of motion to answer the following.
 a) How far will it travel in 30 seconds?
 b) At what speed will it be moving after 30 seconds?

7 A car is travelling along a main road at 14 m/s when it starts to accelerate. Ten seconds later it breaks the speed limit of 18 m/s. By using suitable equations of motion, calculate:
 a) the car's acceleration;
 b) how far it went in the 10 seconds.

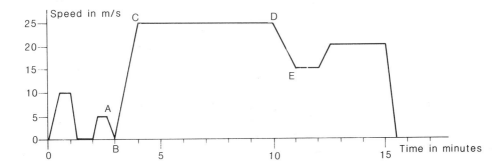

Fig. 2.29

3

THE LAWS OF MOTION

Fig. 3.1

3.1 Introduction

The aeroplane in the photograph has just taken off from a runway. What makes it accelerate along that runway? Why can it not reach a high speed immediately? Why does it reach a maximum speed rather than flying ever faster? If the engines stop working, will it stop flying immediately? If the aeroplane was fitted with twice as many engines, would it have a different acceleration, or a different maximum speed?

To answer questions like these, you need to know about the **laws of motion**. All moving things obey these laws, or *rules*. The laws were discovered by people like Galileo and Newton. This chapter is about those laws.

3.2 You should know . . .

To be able to understand this chapter, you must know and understand:
1 Chapter 2 of this book;
2 Newton's first law of motion.

In case you have forgotten, Newton's first law of motion can be written as follows:

> *If no unbalanced forces push or pull a body, then that body will stay still or keep moving with constant velocity. An unbalanced force will cause a body to accelerate.*

Why has the word 'unbalanced' been used? What does it mean?

3.3   Force and acceleration

You know that you need a force to make something accelerate; you might well ask 'How much force?' This obviously depends on how much acceleration you want. In what way does the acceleration depend on the force? If you double the force, do you double the acceleration, or triple the acceleration, or what?

You may well have investigated this problem before, pulling trolleys with differing numbers of elastics, as in Fig. 3.2. One elastic provides one unit of force, two elastics side by side provide two units of force, and so on.

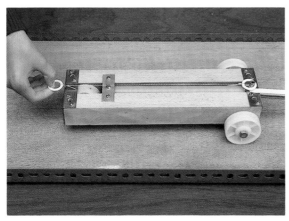

Fig. 3.2   Pulling one trolley with one elastic

If you have never carried out this experiment before, perhaps you could do so now. Pull a trolley with one elastic along a runway, as in the photograph, using a length of ticker tape to record the motion in the usual way. Make a tape chart, or a velocity-time graph, so that you can calculate the acceleration.

Repeat this using two elastics side by side (two units of force) and then three elastics side by side. Calculate the acceleration in each case.

Fig. 3.3 shows the three tape charts Michael obtained when he did this experiment. Do these charts show that a greater force gives a greater acceleration? Do your charts show the same thing?

Do you think that your results of this experiment, or the results in Fig. 3.3, show that if the force is doubled, the acceleration is doubled? In other words, is the acceleration *proportional* to the force?

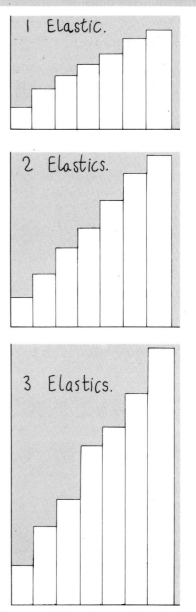

Fig. 3.3   Does a greater force give a greater acceleration?

3.4   Friction

One problem with the experiment with trolleys described in the last section is **friction**; the stretched elastics are not the only force on the trolley. Where in this equipment is there some friction? It is possible to arrange the trolley runway on a *slight* slope which is *just* sufficient to compensate for friction. How do you check that the slope is just right?

There is always friction between moving surfaces. Sometimes it is very small, but it is always there. Although the surfaces may appear absolutely smooth, if you could see them through a powerful microscope they would look something like Fig. 3.4, which shows how the rough surfaces catch each other.

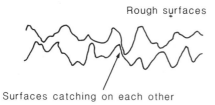

Rough surfaces

Surfaces catching on each other

Fig. 3.4   Rough surfaces create friction

Friction is often a nuisance. If two rough surfaces rub together, a lot of internal energy can be produced—enough to weld the two surfaces together in extreme cases. Friction can be reduced by *lubricating* the surfaces with oil, for example, which helps keep the surfaces apart (Fig. 3.5).

Film of oil keeping rough surfaces apart

Fig. 3.5   Oil can keep rough surfaces apart

Friction is not the only force trying to slow down a vehicle like a car. Another important force is that needed to flatten the tyres slightly where they meet the road—see Fig. 3.6. As the wheel goes round, energy is needed to flatten each part of the tyre as it comes into contact with the road. Where does this energy come from?

Why do you think less energy is needed to move a tonne of goods by rail than a tonne of goods by road?

Friction is not always a nuisance; it is put to use in a car's or motor bike's brakes, for example. The brake pads (Fig. 3.7) rub against a disc attached to the wheels and so provide a force to decelerate the car or motor bike.

Friction is certainly a nuisance in any experiments to investigate the laws of motion, since friction is an *extra* force, and you usually cannot measure the size of it. In accurate experiments we usually try to reduce the friction as much as possible, as in the next experiment.

Fig. 3.6

Fig. 3.7   A disc brake on a motor bike

3.5   Using an air track

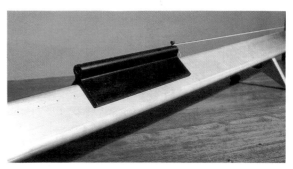

Fig. 3.8   The friction on this air track vehicle is very small

Fig. 3.8 shows a piece of apparatus in which there is very little friction; it is called an **air track**. A vacuum cleaner (which may have to be modified so that it blows instead of sucks), blows air out of the holes in the track. A special vehicle rides on a cushion of air on top of the track; there are no rubbing surfaces and so there is very little friction.

To use this apparatus to investigate the way in which acceleration depends on force, you must decide:

1 how much to push or pull the vehicle with a constant force;
2 how to measure the acceleration.

One possible way of providing a force on the vehicle is to tie a mass to the vehicle and hang the mass over a pulley, as in Fig. 3.9. This mass must be much less than the mass of the vehicle, so that the total mass being accelerated does not significantly change. The mass of the vehicle in the photographs is about 400 grams, while the mass hanging over the pulley is 20 grams. The Earth pulls this mass with a constant force, and it is easy to double the force by doubling the mass.

Do you think this is better than pulling the vehicle with elastics, like the trolley is being pulled in Fig. 3.2? If so, why do you think it is better?

To calculate the acceleration of the vehicle you need to measure the velocity of the vehicle near the start of the track and again further down the track. You also need to measure the time it takes the vehicle to travel between these two points. You can then use the equation

$$\text{acceleration} = \frac{\text{change in velocity}}{\text{time taken}}$$

How can you measure these velocities and this time? Why isn't 'Use a ticker-tape timer' a very sensible answer?

One possible method is to arrange for the vehicle to cut a light beam at both points where the velocity is to be measured, and use some form of light-sensitive sensor to detect when each light beam is broken. This sensor can be connected to a suitable clock which can measure and record when the light beams are broken (see Fig. 3.10).

Fig. 3.11 shows a suitable light-sensitive sensor. It is an example of an electronic **integrated circuit**; it is called a **light-activated switch** (LAS for short). When correctly connected it will switch an electric circuit *on* when light falls on it, and *off* when there is no light on it.

Fig. 3.9   One way to accelerate an air track vehicle

Fig. 3.11   A light-activated switch

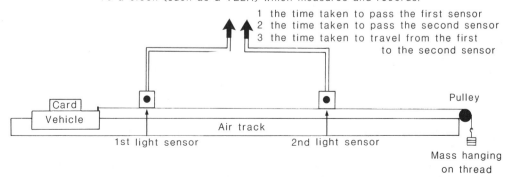

Fig. 3.10   How to measure the acceleration of an air track vehicle

3.6 An electronic clock

The experiment outlined in the previous section needs a suitable clock. There are several possible electronic clocks that can be started and stopped by the light-activated switches. You will see several different examples in this book. What would be particularly useful here is a clock with a memory, so that it can:

1 measure and remember the time taken for the vehicle to pass each light-activated switch, so that you can calculate the two velocities;
2 measure and remember the time taken to travel from one light-activated switch to the other.

A microcomputer has a memory, and it is quite easy to use a microcomputer as a clock which can be controlled by light-activated switches. A microcomputer needs a **program** (that is, a series of instructions) to be written for it. If you know anything about microcomputer programming and have a microcomputer that you can use at school, you might like to try to write a program that uses the microcomputer as a clock controlled by light-activated switches, so that you can measure the acceleration of an air-track vehicle. You should be able to program the microcomputer to do all the necessary calculations for you as well.

Fig. 3.12 shows Vicki using an instrument called a **VELA** as a clock. (A VELA is a Versatile Laboratory instrument—it will do many things; timing events is only one of them. You will see it being used in other ways later in this book.) The VELA has been connected to the light-activated switches which are mounted in the two grey boxes supported on the retort stands nearest to Vicki.

Fig. 3.13   Like much modern equipment, VELA depends on a microprocessor

Like a microcomputer, a VELA has at its heart a microprocessor which will follow a series of instructions (a program). In a VELA, these instructions are stored in an integrated circuit, called a **read only memory** (ROM), so it was not necessary for Vicki to program the VELA before she could use it.

If possible, you should do, or see done, an experiment like the one outlined in Fig. 3.10, using whatever electronic clock you have available in your school. Your teacher will have to show you how to use your particular clock.

3.7 The relationship between acceleration and force is . . . ?

The experiment described in Section 3.5 can be repeated several times, using different masses hanging over the pulley to provide different forces on the vehicle (provided the mass hanging over the pulley

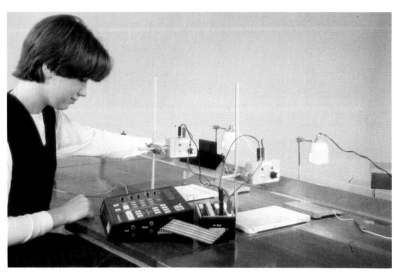

Fig. 3.12   Vicki using a VELA to record times

Results

Length of card on vehicle : 0.2 M

MASS HANGING OVER PULLEY			9	10	20	30	40	50
Time to pass 1st light beam	t_1	S		0.542	0.388	0.313	0.261	0.257
Time to pass 2nd light beam	t_2	S		0.318	0.234	0.185	0.155	0.145
Time to travel from 1st to 2nd light beam	t_3	S		0.928	0.662	0.530	0.462	0.410
Velocity as vehicle passes 1st beam $\left(\frac{0.2m}{t_1}\right)$		M/S		0.37	0.52	0.64	0.77	0.70
Velocity as vehicle passes 2nd beam $\left(\frac{0.2m}{t_2}\right)$		M/S		0.63	0.85	1.08	1.29	1.38
Acceleration = $\left(\text{change in velocity}/t_3\right)$		M/S^2		0.28	0.50	0.83	1.13	1.41

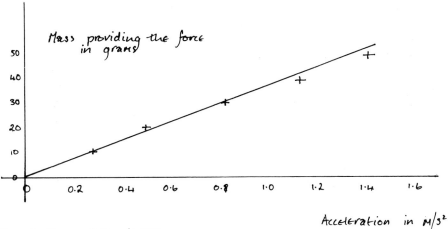

Fig. 3.14   Vicki's graph of force against acceleration

is always much smaller than the mass of the vehicle—why?) The acceleration can be worked out in the same way in each case.

The table above is from Vicki's notebook and shows her results from the experiment you can see her doing in Fig. 3.12. Fig. 3.14 is her graph of these results. Why do you think Vicki drew a graph of the results?

Look at the graph. Notice that Vicki has drawn a straight line, which happens to miss most of the points. Do you think this was the best thing to do? Why not draw a line which joins up all the points?

This graph shows that when Vicki _doubled_ the force pulling the air-track vehicle from the weight of a 10 gram mass to the weight of a 20 gram mass hanging over the pulley, then the acceleration doubled from $0.27\,\text{cm/s}^2$ to $0.54\,\text{cm/s}^2$. Trebling the force (the weight of 10 grams to the weight of

30 grams) trebles the acceleration from $0.27\,\text{cm/s}^2$ to $0.81\,\text{cm/s}^2$.

Have you done an experiment to investigate the relation between force and acceleration? Did you get the same sort of result?

If doubling the force doubles the acceleration, then we say that the acceleration is _directly proportional to_ the force. This conclusion is part of Newton's second law of motion. You will see why it is only _part_ of Newton's second law in the next sections.

A graph of two quantities that are directly proportional to each other will always be a _straight line passing through the origin_, as in Fig. 3.14. This is important. The usual way of finding if two quantities are directly proportional to each other is to see if a graph of the two quantities plotted against each other is a straight line passing through the origin.

3.8 Mass and inertia

Fig. 3.15   Which sledge would *you* rather pull?

Which of the two sledges (Fig. 3.15) would *you* rather be pulling? Although Judith and Ruth are pulling each sledge with much the same force, the acceleration of the sledge that Nicholas is on is much smaller—it is much more reluctant to be accelerated. Once it gets going, of course, it is much more difficult to *decelerate* it to stop it again! We say that the loaded sledge has more **inertia**, or more **mass**. The word inertia comes from the Latin word for laziness. A body with a large mass has a lot of inertia and it is difficult to change its motion.

You can try a simple experiment using two oil cans or similar containers. Fill one of them with water and hang them side by side as in Fig. 3.16.

Fig. 3.16   Appearances may be deceptive!

Try pushing each one to one side. One will obviously be much more difficult to move than the other. Notice that this difficulty has nothing to do with the pull of Earth's gravity on the can; gravity is pulling *downwards* and you are trying to push *sideways*.

You would find exactly the same result even if you were in a spacecraft well away from any noticeable effect of the Earth's gravity.

It is the inertia, or mass, of the aeroplane in Fig. 3.1 that prevents it from starting to fly at a high speed at once. It is its inertia that will keep it moving if the engines stop working.

Can you do this trick? Balance a piece of smooth card (about 5 cm square) on a finger, and put a 50p coin on the card as in Fig. 3.17. Flick the card sharply with the finger of the other hand. The card should fly away leaving the coin balanced on your finger. The inertia of the coin prevents it from accelerating away with the card.

Which coin do you think is the best to use for this trick? Why? Why should the card be smooth rather than rough?

Fig. 3.17
Can you remove the card and leave the coin?

Fig. 3.18
Which piece of cotton will break when you pull?

Try this experiment. Tape two 1 kg masses firmly together and attach them with a length of cotton to a stand as in Fig. 3.18. Tie a length of cotton to the bottom mass as shown. Pull this cotton steadily until one of the pieces of cotton breaks. Which piece breaks?

Repeat the experiment but this time pull the bottom piece of cotton sharply rather than steadily. Does the same piece of cotton break? If not, can you explain why not?

3.9 What is the relation between mass and acceleration?

There are many ways to investigate this problem. For example, you could pull one, then two, then three trolleys with an elastic, measuring the acceleration using a ticker-tape record each time. Or you could carry out an experiment with an air track similar to the one in Section 3.5, pulling vehicles of different masses with the same force each time.

There is yet another way of tackling this problem. Very often physicists make an intelligent *guess* as to what the answer to a problem is going to be and then design an experiment to test whether this guess is right. Often an experiment like this is much simpler and quicker than the sort of experiments outlined in the last paragraph. Of course, if you guess wrong, you still will not know the answer to the problem at the end of the experiment, and you will have to think of another possible answer and another experiment!

Can you do any guessing here? You will probably agree that, if you push with the same force all the time, the *bigger* the mass you are pushing, the *smaller* the acceleration. Could it be that if you *double* the mass then you will *halve* the acceleration (for the same force)? Can you design a *simple* experiment to test if this is true?

Of course, you could do an experiment like one of those outlined above, measuring the accelerations with different masses and checking to see if doubling the mass gives half the acceleration. But measuring accelerations takes time, and it is not really necessary in this case.

Suppose you pull a trolley with one elastic and make a ticker-tape record, as in Fig. 3.2. If you were now to pull two trolleys with one elastic (Fig. 3.19) your guess is that you will have half this acceleration. Suppose you pull two trolleys with *two* elastics (Fig. 3.20). You would now expect two times half the original acceleration (because doubling the force doubles the acceleration). This is the same as the original acceleration! Similarly, three elastics pulling three trolleys would also give the same acceleration (Fig. 3.21). So if your guess is right, you will have the same acceleration if you pull one trolley with one elastic, or two trolleys with two elastics, or three trolleys with three elastics, etc.

Fig. 3.19   Pulling two trolleys with one elastic

Fig. 3.20   Two trolleys and two elastics

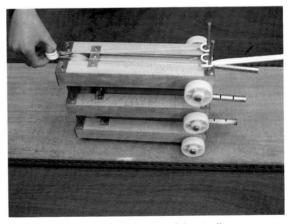

Fig. 3.21   How will the acceleration of these trolleys compare with the previous picture?

This is very easy to check. You do not have to *measure* the acceleration; you just have to put all the pieces of ticker tape side by side and see if they look much the same. This has been done in Fig. 3.22. Do you think these three tapes look very similar? If you carry out this experiment, do you get the same sort of result?

Fig. 3.22 Do these tapes show the same acceleration?

A similar experiment can be done using the air track, light-activated switches and VELA that were used in Section 3.5 earlier in this chapter. All you need to check is that the time taken to go from one light beam to the other is the same in each case.

If the acceleration of anything *halves* when the mass *doubles* then we say that the acceleration is *inversely proportional to* the mass. This is the second part of **Newton's second law of motion.** The complete law now reads:

The acceleration of a body is directly proportional to the unbalanced force applied to the body, and inversely proportional to the mass of the body.

3.10   A unit of force

A kilogram is a scientist's standard unit of mass. A second is the standard unit of time. A metre is the standard of length. What is the standard unit of force and how is it defined?

So far you have used pieces of stretched elastic, or perhaps various weights, to give you the forces you need for your experiments. None of these would make a sensible standard force that people the world over could use. Why not?

Write down one or two reasons why, for example, a piece of stretched elastic would not make a very good 'standard unit of force'.

You know forces cause things to accelerate. Scientists have *agreed* that a sensible standard unit of force is the force that causes a mass of 1 kilogram to accelerate at a rate of 1 m/s². Do *you* think this is sensible? Why?

Two units of force will accelerate a mass of 1 kilogram at 2 m/s², or will accelerate a mass of 2 kilograms at 1 m/s², and so on.

The table below has various combinations of force, mass and acceleration. There is also a fourth column; this is for the value of (mass × acceleration). Copy out the table and fill in the blank spaces.

Force (standard units)	Mass (kg)	Acceleration (m/s²)	Mass × acceleration (kg m/s²)
1	1	1	1
2	1	2	2
2	2		
3	1	3	
4	4		
4		2	
5	5		
8	4		
12	12		
12		4	

Look carefully at your answers. You should be able to see that the figure you put in the fourth column (mass × acceleration) is equal to the force in the first column. So there is a very simple relation between force, mass and acceleration:

$$\text{force} = \text{mass} \times \text{acceleration}$$

Notice that this equation is simple because of the simple way in which scientists have *agreed* to *define* a standard unit for force.

Forces need units. From the equation above you can see that you could use the unit of mass (kg) multiplied by the unit of acceleration (m/s²), and have a force of, for example, 5 kg m/s². Although this is not wrong, scientists usually call the unit of force the **newton**, the symbol for which is N.

Notice that we use a small 'n' for newton when we write out the name in full, but a capital 'N' for the symbol. If a unit is named after a person, the first or only letter of the symbol is a capital one.

We can rewrite the definition of force as:

A force of 1 newton is the force which will cause a mass of 1 kilogram to accelerate at 1 m/s².

3.11 Using the equation: force = mass × acceleration

We usually use the letter F to stand for force, m for mass and a for acceleration, so the equation in the heading to this section would usually be shortened to

$$F = ma$$

This section has two simple examples of how you use this equation to help solve problems.

Example 1

The sledge Nicholas is on in Fig. 3.15 has a total mass of 30 kg.
a) With what unbalanced force must Judith pull the sledge so that it accelerates at 0.5 m/s²?
b) In fact, Judith would have to pull with a bigger force than you have just calculated. Why?

Part a)

$$\text{mass} \quad m = 30\,\text{kg}$$
$$\text{acceleration} \quad a = 0.5\,\text{m/s}^2$$

Using
$$F = ma$$
$$F = 30\,\text{kg} \times 0.5\,\text{m/s}^2$$
$$= 15\,\text{N}$$

Judith must pull with a force of 15 N.

Part b)
As well as supplying a force of 15 N to accelerate the sledge, Judith must supply *extra* force to pull against friction and provide the force needed to compress the snow in front of the sledge. The 15 N calculated in part a) is the *unbalanced* (or 'resultant') force needed to produce an acceleration of 0.5 m/s².

Example 2

A boy pushes a go-cart with a force of 40 N. The mass of the go-cart is 20 kg.
a) What will be the acceleration of the go-cart?
b) If the boy keeps pushing with this force, what will be the velocity of the go-cart after 10 seconds, assuming he starts from rest?
c) Do you think this answer is reasonable?

Part a)

$$\text{mass} \quad m = 20\,\text{kg}$$
$$\text{force} \quad F = 40\,\text{N}$$

Using
$$F = ma$$
$$40\,\text{N} = 20\,\text{kg} \times a$$
$$a = \frac{40\,\text{N}}{20\,\text{kg}}$$
$$= 2\,\text{m/s}^2$$

The acceleration of the trolley is 2 m/s².

Part b)
This part is nothing to do with $F = ma$, but is solved using the equations of motion which were introduced in Sections 2.9 to 2.11.

$$\begin{aligned}
\text{initial velocity} \quad & u = 0\,\text{m/s} && \text{(Starts from rest.)} \\
\text{acceleration} \quad & a = 2\,\text{m/s}^2 && \text{(We have just calculated this.)} \\
\text{time} \quad & t = 10\,\text{s} && \text{(See question.)} \\
\text{final velocity} \quad & v = ? && \text{(This is what we are being asked to find.)}
\end{aligned}$$

Using
$$v = u + at$$
$$v = 0 + 2\,\text{m/s}^2 \times 10\,\text{s}$$
$$= 20\,\text{m/s}.$$

After 10 seconds, the boy will be pushing the go-cart at 20 m/s.

Part c)
(i) There will be some friction acting against the force of 40 N with which the boy pushes, so not all the 40 N will be available to cause acceleration, so the acceleration will be less than 2 m/s².
(ii) It takes an Olympic sprinter about 10 s to run 100 m (average speed of 10 m/s) and such a sprinter does not push a go-cart! Our calculation shows that the boy will be moving at twice this speed after 10 s! Is that reasonable?
There is obviously a top speed at which anybody can run, and that top speed is much less than 20 m/s.

29

SUMMARY

Now that you have finished studying this chapter on Newton's laws of motion, there are a number of things you should know and be able to do.

1 You should:
   a) know Newton's first law of motion;
   b) know Newton's second law of motion;
   c) know what is meant by 'directly proportional';
   d) know what is meant by 'inversely proportional';
   e) know that if two variables are directly proportional to each other, then a graph of one against the other will be a straight line passing through the origin;
   f) be able to describe at least one experiment which shows that acceleration is proportional to resultant force;
   g) be able to describe at least one experiment which shows that acceleration is inversely proportional to mass;
   h) be able to use the equation $F = ma$;
   i) know the definition of 1 newton of force;
   j) know that there is friction between moving surfaces, and that friction provides a 'retarding' force.

2 You should know what each of the following is, or what each does:

friction	brake pads
air-track	sensor
light-activated switch	inertia
mass	instrument
origin (of graph)	definition
resultant force	retarding force
unbalanced force	conclusion
microprocessor	

3 Here are some of the other words that have been used in this chapter. You should know what each word means:

compensate	investigate
device	succession
program	modify
restore	display
instruction	reduce
reluctant	standard
lubricate	

FURTHER QUESTIONS

1 In the introduction at the start of this chapter, a number of questions were asked about an aeroplane. As a result of studying this chapter, how many of those questions can you now answer, and what are the answers?

2 Section 3.5 described an experiment using an air track. What advantages can you think of in doing the experiment in the way described rather than using trolleys and ticker tape. What disadvantages does this method have?

3 In Fig. 2.1 on page 8 the engine of the high-speed train was pulling hard, but the train was moving at a constant speed. In this chapter you have learned that a force produces an acceleration. Why did the train not accelerate?

4 Fig. 3.23 shows five graphs, labelled A to E, but without the axes marked. Which graph (or graphs) could be:

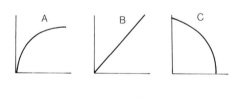

Fig. 3.23

a) a velocity-time graph of something moving with constant acceleration?
b) a graph of the force applied to an air-track vehicle against the acceleration of the vehicle?

c) a distance-time graph of something moving with constant velocity?

d) a distance-time graph of a train coming to rest at a station?

e) a graph which shows direct proportionality?

f) a distance-time graph for something moving with constant acceleration?

g) a graph of acceleration against time for a frictionless trolley being pushed with a constant force?

5 A school bag full of books has a total mass of 10 kg. A force of 20 N is needed to make it slide along a bench at a constant velocity.

a) What is the unbalanced (or resultant) force acting on the bag when the total force is 30 N?

b) What will be the acceleration of the bag when the total force acting on the bag is 30 N?

c) What is the total force with which the bag must be pulled in order to double the acceleration?

6 A lorry has mass of 5 tonnes (5000 kg). It accelerates steadily from rest to 30 m/s in 10 s.

a) What is the lorry's acceleration?

b) What force is needed to produce this acceleration?

c) How far will the lorry travel in 10 s?

d) The actual force exerted by the engine will be greater than you have calculated in part b). Why is this?

7 A 1200 kg car is travelling at 15 m/s. It crashes into a wall and stops in 0.10 s.

a) What is the change in the car's velocity as it stops?

b) What is the deceleration of the car?

c) What is the average force stopping the car in that time?

d) Why is this an *average* force?

8 A tractor is towing a trailer of mass 500 kg at a steady speed along a level road. The tension (pulling force) in the tow bar is 300 N.

a) Should not the tension in the tow bar be zero if the trailer is moving at constant speed? Explain your answer.

b) The tension now increases to 1000 N. What is the acceleration of the trailer?

c) What is the acceleration of the tractor?

9 Imagine that you are standing in a bus which stops suddenly. You fall forwards. Is it true to say that the bus has 'thrown you forwards'? If not, what would be a better explanation of why you fall forwards?

10 The Greek philosopher Aristotle taught his students that a constant force was required to produce a constant velocity and from this he concluded that, in the absence of any force acting on it, a body would come to rest.

a) What is the difference between Aristotle's ideas and Newton's laws of motion?

b) Write down several situations where a constant force seems to produce a constant velocity.

c) How do you explain each of your situations using Newton's laws of motion?

4

FALLING

4.1 Introduction

Fig. 4.1   What is her acceleration?

Why is this diver falling? What pulls her down? What a silly question, you might think! Gravity pulls her down! But it is not at all obvious *why* the Earth should pull everything towards it, is it?

There are other questions we could ask about gravity. For example, how *big* is the force of gravity on the diver? This force makes the diver accelerate down towards the water; what is the acceleration of the diver?

When you have finished reading this chapter, you should be able to answer these questions.

4.2 You should know . . .

To be able to understand this chapter, you should know and understand all that is in Chapters 2 and 3 of this book. You will also find it helpful to know a little about stroboscopic photography, but it does not matter if you have not met this technique before, since you will find that Section 4.4 contains all the necessary details.

4.3 The acceleration due to gravity

What *was* the acceleration of the diver? Was it the same as anything else dropping towards the Earth, or does everything have its own acceleration?

Galileo was a famous Italian scientist who lived in the sixteenth century. There is a story that one day he dropped a musket ball and a cannon ball from the top of the Leaning Tower of Pisa. The onlookers expected the cannon ball to arrive at the

Fig. 4.2   The site of Galileo's experiment

ground first. They were wrong. Both balls arrived together. Galileo was trying to show that *everything* falls to the Earth with the same acceleration.

Unfortunately, the story is not quite true. As far as we know Galileo never carried out this experiment, but he did *think* about what would happen in such an experiment.

Nevertheless, the conclusion is correct. Everything accelerates towards the Earth at the same rate.

But surely all this is nonsense, you might be thinking. How can the acceleration of free-fall be the same for everything? If you drop a feather and a golf ball, they do not arrive at the ground at the same time! This is true, but it is because of **air resistance**. Air resistance has a noticeable effect on the feather because it is so light. There is air resistance acting on the golf ball but the effect is negligible because the golf ball is heavy and compact.

If you were to drop a feather in a vacuum, where there is no air, it would accelerate at just the same rate as the golf ball. On one of the Apollo space flights to the Moon (where there is no air) David Scott dropped a hammer and a feather at the same time, for the benefit of television viewers at home, to see if they would hit the Moon's surface together—they did.

4.4 Measuring the acceleration due to gravity using stroboscopic photography

Since everything accelerates towards the Earth at the same rate, there is no need actually to measure the acceleration of the diver in Fig. 4.1 to answer one of the questions in the first section. Designing an experiment to measure the acceleration of, say, a falling golf ball in a laboratory will be much easier and the answer will be exactly the same (provided air resistance can be neglected).

Fig. 4.4 suggests one way of measuring the acceleration of a falling golf ball. The ticker-tape record can be used for making a velocity-time graph, from which you can measure the acceleration—how? Try this if you wish. Although you will probably obtain an answer, it won't be a very reliable one. This is not a very good way of solving the problem.

Write down what you think is wrong with this experiment. (You may disagree and think this is a very good way! Why do you think it is good?)

Fig. 4.3 There is no atmosphere on the Moon—so there is no air resistance!

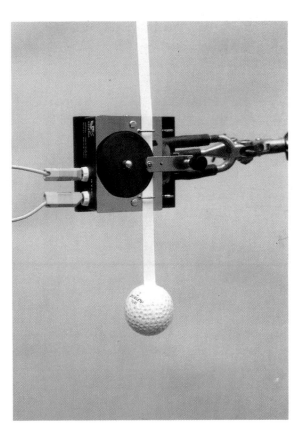

Fig. 4.4 Does this look like a sensible experiment?

33

One very useful way to make a record of something moving is to use **stroboscopic photography**. Fig. 4.5 shows how to arrange equipment to take a stroboscopic photograph of a falling golf ball. The laboratory should be as dark as possible, with a black surface (like a curtain) behind the falling ball.

In Fig. 4.5 Kate was in charge of the camera. She used a fast film (Ilford HP5) with a wide camera aperture (f/2). The shutter was open all the time the release was pressed, that is, it was on the 'B' setting. When she gave the word she opened the shutter (using the cable release), and at the same time Michael dropped the ball in front of the scale, which was marked in 10 cm intervals. Kate let the shutter close when the ball hit the ground. The stroboscope (visible on the left of the photograph) was flashing 25 times per second.

Fig. 4.6 is one of the photographs that Kate and Michael obtained. It looks similar to a short length of ticker tape. Fig. 4.7 shows you how to analyse this photograph to obtain figures for a velocity-time graph. Fig. 4.8 is the graph that Kate obtained from these results. What is the acceleration shown by this graph?

The last photograph in this section shows a stroboscopic photograph of a golf ball and ping-pong ball falling together. What do you notice about the acceleration of the two balls?

Fig. 4.5 Taking a stroboscopic photograph

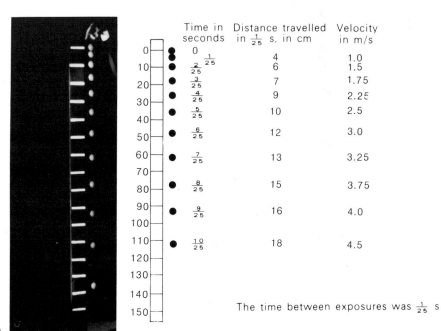

Fig. 4.6
A strobe photograph of a falling ball

Time in seconds	Distance travelled in $\frac{1}{25}$ s, in cm	Velocity in m/s
0		
$\frac{1}{25}$	4	1.0
$\frac{2}{25}$	6	1.5
$\frac{3}{25}$	7	1.75
$\frac{4}{25}$	9	2.25
$\frac{5}{25}$	10	2.5
$\frac{6}{25}$	12	3.0
$\frac{7}{25}$	13	3.25
$\frac{8}{25}$	15	3.75
$\frac{9}{25}$	16	4.0
$\frac{10}{25}$	18	4.5

The time between exposures was $\frac{1}{25}$ s

Fig. 4.7

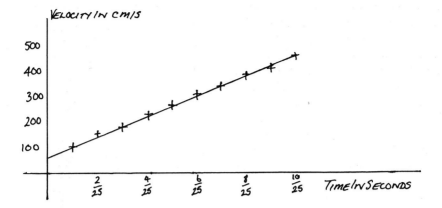

Fig. 4.8 The velocity-time graph of the falling ball

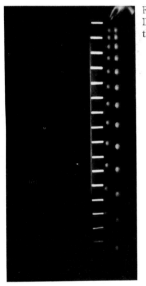

Fig. 4.9
Do both balls accelerate at
the same rate?

The idea behind the experiment is very simple. You drop the golf ball from a known height and measure the time it takes to reach the ground—see Fig. 4.10. You can then use the equation of motion

$$s = ut + \tfrac{1}{2}at^2$$

to calculate the acceleration a. You can measure the distance s the ball falls (Using what?) and also measure the time taken to fall using a suitable clock. In this case $u = 0$. Why?

4.5 Measuring the acceleration due to gravity using an electronic timer

Although stroboscopic photography is a very useful way of recording the way something moves, especially complicated movements, you will probably agree that it takes quite a long time to carry out the experiment and calculate any results. In this section you will learn of a quicker way of finding the acceleration due to gravity.

This method does not allow you to check that the acceleration does not change as the golf ball falls; that has to be checked using the stroboscopic photography method. What is it about the graph in Fig. 4.8 that shows that the acceleration of the golf ball was indeed constant?

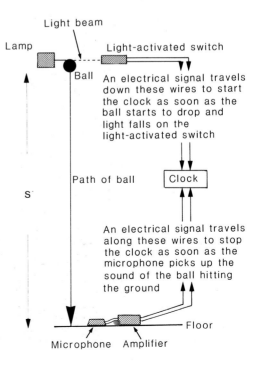

Fig. 4.10 Another way to find the acceleration of free fall

The main problem, as usual, is finding a suitable clock. It must be possible to start and stop it automatically at the right moments (Why?) and have a suitable sensitivity. (What does this mean?)

You could use a VELA, or you could use another electronic clock like the **scaler-timer** in Fig. 4.11. This will measure times in milliseconds. What fraction of a second is a millisecond?

Fig. 4.11   A scaler-timer

The timer can be started or stopped using the switches at the bottom of the instrument. It can also be started using the 'start' terminals. The clock will start as soon as the terminals are connected together with a length of wire, or, if they are already connected together, as soon as they are disconnected. You can also connect a device like the light-activated switch that you learned about in Section 3.5 to these terminals. The clock will start as soon as the output from the light-activated switch changes from 'low' to 'high', that is, as soon as light falls on the light-activated switches. The 'stop' terminals work in a similar way to stop the clock.

(*Note* Not all timer-scalers work in quite the same way as this model.)

Fig. 4.12 shows Sheila and Richard carrying out this experiment. Sheila is holding a ball in front of the light-activated switch mounted at the top of the retort stand rod. This light-activated switch is connected to the 'start' terminals of the timer-scaler. As soon as she releases the ball, light will fall on the light-activated switch and the timer will start.

Sheila and Richard have placed a microphone on the ground to provide the 'stop' signal to the timer. The microphone is connected, via a small amplifier, to the 'stop' terminals. As soon as the ball hits the ground, the microphone picks up the sound of it doing so, and hence stops the clock. The distance that the ball fell in this experiment was 1.5 m.

Fig. 4.13 shows the times (in milliseconds) that the timer measured on three different occasions that they carried out this experiment. What was the *average* time?

Use these measurements in the equation

$$s = ut + \tfrac{1}{2}at^2$$

to calculate the acceleration of the ball. Does this acceleration agree with the acceleration shown by the stroboscopic photography method in the last section?

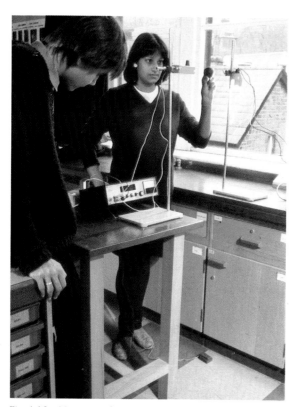

Fig. 4.12   Measuring the acceleration of free fall

Fig. 4.13   Three timer readings from the previous experiment

4.6 Gravitational field strength

After many different kinds of experiments and measurements, physicists agree that the acceleration due to the Earth's gravity is about 9.8 m/s². From the experiments you have done or seen, do you agree? It varies a little depending where on the Earth you are. For most purposes it is convenient to call this acceleration 10 m/s².

Fig. 4.14 What is the *weight* of this sugar?

What is the force with which the Earth pulls the bag of sugar? This force is called the **weight** of the bag of sugar. Its weight is *not* 1 kilogram—that is its **mass**. If you were to drop the bag it would accelerate down at 10 m/s². The force (its weight) causing this acceleration can be calculated using the equation $F = ma$.

$$F = ma$$
$$F = 1 \, kg \times 10 \, m/s^2$$
$$= 10 \, N$$

So the weight of the bag of sugar is 10 N. The words 'mass' and 'weight' are often used wrongly in everyday life. When shopping, for example, you will talk about weight when really you mean mass.

A similar calculation to the one above shows that the weight of a 2 kg mass is 20 N, a 3 kg mass weighs 30 N, and so on. For every kilogram of mass the Earth pulls down with a force of 10 N in order to cause an acceleration of 10 m/s². We say that the **gravitational field strength** of the Earth is 10 N/kg.

The diver in Fig. 4.1 had a mass of 45 kg, what was her weight? What is your weight?

Astronauts on the Moon have much less weight than on the Earth. The astronaut's *mass* is the same on the Moon as on the Earth (after all, there is the same 'amount of astronaut' in both places), but the gravitational field strength of the moon is only 1.6 N/kg, so each kilogram of astronaut is pulled towards the Moon with roughly $\frac{1}{6}$ of the force on the Earth.

What would the diver's weight be on the Moon? What would your weight be?

Large planets have a much bigger gravitational field strength; on Jupiter (the largest planet in the solar system) it is 25.4 N/kg. How much would you weigh on Jupiter?

4.7 Terminal velocity

Fig. 4.15 Why doesn't this parachutist keep accelerating?

In the last three sections, have we not forgotten the golf ball and feather experiment of Section 4.3? Surely the parachutist in the photograph is not accelerating at 10 m/s²? The whole point of a parachute is to make you come down at a steady, and relatively slow, speed.

The things we have been dropping in recent sections have been compact and dense, so that air resistance has been a comparatively small force. With a parachute there is a lot of air resistance acting against gravity.

When the parachutist jumped, he accelerated. The faster the parachutist moved, the greater was the force provided by the air resistance; eventually the air resistance was equal to the parachutist's weight. Then he descended with constant velocity since there was no unbalanced force acting on him. This constant velocity is called the parachutist's **terminal velocity**.

The fact that 'the faster you move, the greater the wind resistance' is important for the designers of vehicles that need to move fast. Why, for example, is the front of the high-speed train (Fig. 2.1) shaped the way it is? And look at the fuel consumption figures for a car in Fig. 4.16; all that extra fuel at the higher speed is because the engine has to work harder to provide the force to overcome the extra wind resistance.

Economy					
	Imperial mpg			Metric L/100km		
	Urban	56 mph	75 mph	Urban	90 km/h	120 km/h
Manual	39.6	60.5	41.5	7.1	4.7	6.8

All figures (in mpg – L/100km) are from officially approved tests under the Passenger Car Fuel Consumption Order 1977.

Fig. 4.16   Fuel consumption figures for an Austin Maestro. Urban figures refer to 'stop-start' driving. Why is this figure so low?

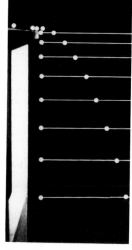
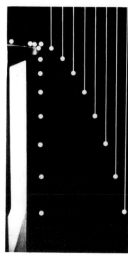

Fig. 4.18
One ball fell vertically, one fell 'sideways'

Fig. 4.19

4.8   Falling sideways

One thing we have so far overlooked about the diver in Fig. 4.1 is that she is not falling straight down; she is moving slightly to the right as well as moving down. Does this make any difference to her acceleration?

You can examine this problem by taking a stroboscopic photograph of two golf balls falling together, one falling straight down and the other also moving sideways. Fig. 4.17 shows a way of launching the two balls together: one ball rolls down the ramp and hits the ball at the bottom; this ball falls outwards, while the first ball falls straight down. Fig. 4.18 shows the stroboscopic photograph of this motion. What do the horizontal lines drawn on the photograph tell you about the downward accelerations of the two balls?

Fig. 4.19 is the same photograph, but with vertical lines drawn through the positions of the ball moving sideways, showing how far sideways it went between each flash of the stroboscope. What does this picture tell you about the horizontal part (or *component*) of the velocity?

The next photograph is of a ball launched upwards; the shape of the path it makes is called a **parabola**. The curved paths in Figs 4.18 and 4.19 are also parabolas. What do you notice about the upwards deceleration and the downwards acceleration after reaching the top of its flight? What about the horizontal velocity?

Fig. 4.17   How to launch two balls

Fig. 4.20   This ball was fired upwards

Copy out and complete the following paragraph, choosing your words or phrases from the list at the end. You will not need all the words.

'When something, which is moving in a vertical gravitational field, has a horizontal _____ to its velocity, it will move with _____ horizontally and _____ vertically. The shape of the path it makes is called a _____.'

a constant acceleration   a constant velocity
zero velocity   parabola   straight line
component   a force   a downward force

4.9   Where does the stunt-man hit the water?

Here is a numerical example of 'falling sideways'.

In a film, a stunt-man has to dive off a cliff 20 m high, into some water. He dives in such a way that his horizontal velocity is 6 m/s and his vertical velocity is zero. How far out from the edge of the cliff does he hit the water?

To solve this problem, you must remember that the horizontal velocity will stay the same (6 m/s) all the time, and that he will accelerate in a vertical direction at 10 m/s².

First, calculate how much time it takes to fall vertically into the water:

initial velocity $u = 0$ (This is the initial velocity vertically downwards.)

distance $s = 20$ m
acceleration $a = 10$ m/s²   (Due to gravity.)
time $t = ?$

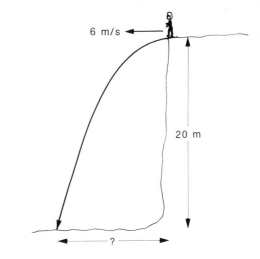

6 m/s

20 m

?

Fig. 4.21

Using

$$s = ut + \tfrac{1}{2}at^2$$
$$20\,\text{m} = 0 + \tfrac{1}{2} \times 10\,\text{m/s}^2 \times t^2$$
$$t^2 = 4\,(\text{s})^2$$
$$t = 2\,\text{s}$$

It takes 2 seconds to fall from the top of the cliff into the water. During this 2 seconds, the stunt-man is moving horizontally at 6 m/s; he will travel 12 m horizontally in 2 seconds, so he will land in the water 12 m from the cliff.

(This calculation ignores the effect of wind resistance.)

SUMMARY

Now that you have finished studying this chapter on falling, there are a number of things you should know and be able to do.

1 You should:
   a) understand the difference between mass and weight;
   b) know that the acceleration due to gravity is the same for all objects, provided you can neglect the effect of air resistance;

   c) know what is meant by 'terminal velocity', and why anything falling through the air has a terminal velocity;
   d) be able to measure the acceleration of free-fall by at least one method;
   e) understand what is meant by 'gravitational field strength';
   f) know what a stroboscopic photograph is, and how to make one;

g) understand why it is less economical on fuel to travel fast rather than slow;

h) understand the way in which something falls if it is also moving sideways.

2 You should know what each of the following is, or what it does:

stroboscope scaler-timer
component parabola
 (of velocity)
shutter (of camera)

3 Here are some of the other words that have been used in this chapter. You should know what each word means:

resistance reliable
recording instrument
automatically overlook
convenient astronaut
descend consumption

FURTHER QUESTIONS

In all these questions, assume that the acceleration due to the Earth's gravitational field is $10\,m/s^2$.

1 If a mouse falls $10\,m$ from the top of a house, it will probably just run away unhurt after it lands. If you fall the same distance, you will almost certainly break a few bones, if nothing worse. Why is there this difference? Surely both you and the mouse hit the ground at the same velocity?

2 A space traveller weighs $850\,N$ on Earth.
a) What is the mass of the traveller?
b) What is the weight of the traveller on a smaller planet, where the gravitational field strength is $3\,N/kg$?
c) On a larger planet the traveller weighs $1360\,N$. What is the gravitational field strength on this planet?

3 A helicopter hovers $45\,m$ above some snow-covered fields to drop hay to some stranded cattle. How long does it take a bale of hay to drop to the ground?
What have you had to ignore in answering this question?

4 A cricketer hits a cricket ball vertically upwards at a velocity of $20\,m/s$.
a) How high does the ball go?
b) How much time elapses before the ball hits the ground?

5 A passenger in a moving train accidentally drops a book out of the window. He watches it drop to the ground and decides that it dropped straight down. However, somebody by the side of the track watching the train pass by says that the book fell in a curved path towards the ground.
a) Make a sketch showing the curved path of the falling book as seen by the person at the side of the track.
b) Explain why the two people disagree about the way in which the book fell to the ground.

6 a) A rocket on a launching pad has a total mass of $1200\,kg$. What is the rocket's weight?
b) The rocket's engines push with a force (thrust) of $15\,000\,N$. What is the resultant (un-balanced) force on the rocket?
c) What is the acceleration of the rocket?

7 A survey aircraft is flying at $300\,m/s$, $80\,m$ above the sea when it releases a packet of scientific equipment.
a) What is the initial downward velocity of this equipment?
b) How long does the packet take to reach the sea? (Ignore air resistance.)
c) What is the horizontal distance between the spot where the equipment was released and the spot where it hits the sea?

5

COLLISIONS

Fig. 5.1 Collisions!

5.1 Introduction

It is unlikely that the people enjoying themselves on the 'bumper' cars were giving much thought to physics, but physicists are very interested in the rules about the way in which things bump into each other. One reason for this is because an important method of finding out about how atoms are made is to 'fire' high-speed sub-atomic 'missiles' at atoms and to look at what happens during the collision.

What do you think 'sub-atomic' means?

Why is there a thick strip of rubber round the bottom of the bumper cars in the photograph? In what way would the collisions between cars be different without this rubber strip? Some real cars have very similar thick rubber bumpers. It would be a sensible guess that this somehow makes them safer. Why does it make them safer? This chapter is about finding the answers to questions like these.

Fig. 5.2 Why does it have a thick rubber bumper?

5.2 You need to know . . .

In order to understand this chapter, you must have read and learned the physics in Chapters 2, 3 and 4 of this book.

5.3 Trying to stop things moving

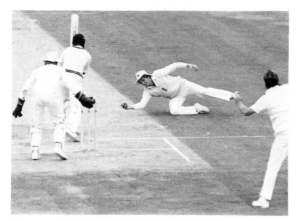

Fig. 5.3

Fig. 5.5   Heavily used motorways have crash barriers

Which would you rather catch, a golf ball coming towards you at 5 m/s, or a cricket ball coming at the same speed? Most people will prefer the golf ball, which has a smaller mass than a cricket ball; it is easier to stop and is less likely to hurt them.

Fig. 5.4   Which would you rather catch?

Suppose now the choice is between a cricket ball travelling at 5 m/s and a cricket ball travelling at 10 m/s? Most people will prefer to catch the slower moving cricket ball; it is easier to stop and is less likely to hurt them.

The faster a ball is moving, and the greater the ball's mass, the more difficult it is to stop and the more likely it is to hurt you.

Which is going to be more difficult for a motorway crash barrier to stop, a car which bumps into it at 15 m/s, or the same car bumping into it at 30 m/s? Which will cause more damage to the crash barrier?

Which is more difficult for a crash barrier to stop, a car moving at 30 m/s, or a lorry moving at 30 m/s?

The fast car is more difficult to stop than the slow car because it has more **momentum**. The fast moving cricket ball has more momentum than the slow moving cricket ball. The slow moving cricket ball has more momentum than the slow moving golf ball. The more momentum something has, the more difficult it is to stop and the more damage it can do or the more likely it is to hurt you.

The momentum of anything is calculated by multiplying its mass by its velocity:

$$\text{momentum} = \text{mass} \times \text{velocity}$$

The units of momentum are kg m/s (or kg m s^{-1}).

If the golf ball travelling at 5 m/s has a mass of 50 grams (0.05 kg), then its momentum is calculated from

$$\begin{aligned}\text{momentum} &= \text{mass} \times \text{velocity} \\ &= 0.05 \text{ kg} \times 5 \text{ m/s} \\ &= 0.25 \text{ kg m/s}\end{aligned}$$

The momentum of the golf ball is 0.25 kg m/s.

What is the momentum of the cricket ball (mass 150 grams) when it is moving at:
a)   5 m/s;
b)   10 m/s?

If the car mentioned earlier in this section has a mass of 1 tonne (1000 kg) and the lorry has a mass of 5 tonnes, what is the momentum of each vehicle when travelling at:
a)   15 m/s;
b)   30 m/s?

5.4 Trying not to get hurt

You do not *have* to hurt your hand when you catch a ball. If you let your hand move along with the ball as you catch it, you are less likely to hurt yourself. If you catch a ball in this way, you take a little longer to stop the ball. The deceleration of the ball is therefore smaller so less force (provided by your hand) is needed to cause this deceleration.

There is another way of looking at this. Before the ball lands in your hand it has some momentum. After it has been caught it has no momentum because it has stopped. It has lost its momentum because your hands provided a force to stop the ball. To change the ball's momentum, you can either provide a *large* stopping force for a *short* time (your hands do not move with the ball) or a *smaller* stopping force for a *longer* time (your hands do move with the ball).

Here is another example of the same idea. Suppose you jump from a wall to the ground (Fig. 5.6). When you land, you probably bend your legs as Toby is doing in the photograph. Why? What happens if you do not bend your legs?

Fig. 5.7   Why is there a mattress to land on?

Fig. 5.6   Why does Toby bend his legs as he lands?

Just before you land on the ground you have some momentum. After you have landed you have no momentum. If you do not bend your legs you will stop in a very short time and so change your momentum in a very short time. This needs a large force pushing on you to stop you; you will hurt yourself. If you do bend your legs you take a longer time to stop and this requires less force; you do not hurt yourself.

Use the same idea to explain why there is a mattress for the gymnast to land on in Fig. 5.7. What would happen if she landed on the floor?

We have talked about *force* and *time* in this section. A large force acting for a small time can give exactly the same momentum change as a small force acting for a long time. Physicists say that both can provide the same **impulse**. Impulse is calculated by multiplying the force by the time for which it is applied:

$$impulse = force \times time$$

The units of impulse are N s (newton seconds).

A force of 8 N acting for 5 seconds gives an impulse of $8 N \times 5 s = 40 N s$. This is exactly the same impulse as 4 N acting for 10 s, or 2 N acting for 20 s, or 1 N acting for 40 s, and so on.

Now explain why there are rubber bumpers round the bumper car in Fig. 5.1.

Modern cars are designed so that, if they crash into something, the front end (the bonnet and front wings) crumples. The car is then brought to a more gradual stop than if the front end were rigid. The passenger compartment, however, is made as rigid as possible to protect the people in the car.

Fig. 5.8   Modern car wings and bonnets are designed to crumple like this. Why?

43

5.5 The relationship between impulse and momentum

In the last section you saw that you can change the *momentum* of anything by applying an *impulse*. This section is about finding a simple mathematical relationship between momentum and impulse.

Our starting point is the equation

$$\text{force} = \text{mass} \times \text{acceleration}$$

or $\qquad F = ma$

The acceleration a is the

$$\frac{\text{change of velocity}}{\text{time}}$$

so we can rewrite this equation as

$$F = m\frac{(v-u)}{t}$$

(The letters have their usual meanings.) Multiplying both sides of the equation by t:

$$Ft = mv - mu$$

Look at the different parts of this equation.

1 On the left-hand side is Ft, force multiplied by time. This is what we called an 'impulse' in the last section.
2 On the right-hand side is mv, mass multiplied by final velocity. Mass multiplied by velocity is what we have called 'momentum', so this term must be the 'final momentum', that is, the momentum after the impulse has finished.
3 Also on the right-hand side is mu. This is the 'initial momentum', the momentum before the impulse is applied.

The whole of the right side of this equation is the *change* in momentum.

We can write this equation in words as:

An impulse results in a change in momentum.

This, of course, is what we agreed at the end of the last section, but this equation also tells us that the impulse is *numerically equal* to the momentum change.

Since an impulse is equal to the momentum change it produces, the units of impulse and momentum must be the same as each other. It is perfectly correct to measure momentum in N s rather than kg m/s if you wish.

5.6 An example using the relationship 'impulse = change in momentum'

What is the average force with which a footballer kicks a football from rest to make it move with a velocity of 30 m/s? The football has a mass of 0.5 kg, and the kick lasts 0.02 s.

$$\begin{aligned}
\text{mass} \quad & m = 0.5\,\text{kg} \\
\text{initial velocity} \quad & u = 0 \\
\text{final velocity} \quad & v = 30\,\text{m/s} \\
\text{time} \quad & t = 0.02\,\text{s}
\end{aligned}$$

Using
$$Ft = mv - mu$$
$$F \times 0.02\,\text{s} = 0.5\,\text{kg} \times 30\,\text{m/s} - 0$$
$$F = \frac{0.5\,\text{kg} \times 30\,\text{m/s}}{0.02\,\text{s}}$$
$$= 750\,\text{N}$$

The footballer kicks the ball with an average force of 750 N.

5.7 Why wear a seat belt in a car?

In Britain, and in many other countries, you must wear a seat belt if you are travelling in the front seat of a car. We can use these ideas of impulse and momentum to understand why you will be hurt less in a crash if you wear a seat belt.

Fig. 5.9   Why is it sensible to wear a seat belt?

If a car stops suddenly (because it collides with something) and you are not wearing a seat belt, you will keep moving with whatever momentum you had before the car stopped, until you hit the dashboard. You then lose all your momentum in a very short time, and this requires a very large force.

Imagine a car, with you in it, is moving at 15 m/s when it has a collision. Suppose you have a mass of 60 kg and when you hit the dashboard you stop in $\frac{1}{50}$ s. What force acts on you to stop you?

$$
\begin{aligned}
\text{mass} \quad m &= 60 \, \text{kg} \\
\text{initial velocity} \quad u &= 15 \, \text{m/s} \\
\text{final velocity} \quad v &= 0 \, \text{m/s} \\
\text{time} \quad t &= 0.02 \, \text{s}
\end{aligned}
$$

Using   'impulse = change in momentum'
$$Ft = mv - mu$$
$$F \times 0.02 \, \text{s} = 0 - 60 \, \text{kg} \times 15 \, \text{m/s}$$
$$F = -\frac{60 \, \text{kg} \times 15 \, \text{m/s}}{0.02 \, \text{s}}$$
$$= -45\,000 \, \text{N}$$

The force acting on you will be 45 000 N.

This is equal to the weight of 4.5 tonnes, roughly the weight of three large Land Rovers. Would you like three large Land Rovers on top of you?

Suppose you did wear a seat belt. If a car stops suddenly, the seat belt stretches and brings you to a more gradual stop; you take longer to lose your momentum and so the force on you is smaller than before. Suppose you now take $\frac{1}{5}$ s to stop. This is *ten times longer* than when you were not wearing a belt, so the force on you will be *ten times smaller*. The force on you this time will be 4500 N, or roughly the weight of half a tonne. This is the weight of about a dozen of your friends. A dozen people piled on top of you would be uncomfortable (Have you ever been at the bottom of a rugby scrum?) but you certainly will not come to as much harm as with 4.5 tonnes on top of you.

Fig. 5.10   Would you like to be at the bottom of this?

5.8   Newton's third law of motion

Fig. 5.11   Who is pulling whom?

Who is pulling whom in Fig. 5.11? Of course, they are both pulling each other. Since the tension in the rope is the same all the way along, they must be pulling each other with equal, (but opposite), forces. This serves as an illustration of **Newton's third law of motion**, which says:

For every force, there is an equal and opposite force (or 'reaction', as it is sometimes called).

Notice that the law says nothing about *where* the equal and opposite force is to be found. A force and its reaction always act on *different* bodies. If they always acted on the same body, nothing could ever accelerate—why not?

If you are sitting on a chair, you push down on the *chair* with a force equal to your weight and the chair pushes up on *you* with an equal force, but in the opposite direction. But what forces are there if the chair is pulled away from under you? There is a force pulling you down (your weight). Where is the equal and opposite force?

The answer may surprise you. You pull *up* on the Earth with an equal force. The effect on the Earth is negligible, because the Earth is so massive, so do not expect the Earth to come up to meet you next time you fall from a chair.

Newton's law of gravitation (which you will learn more about later in this book), is concerned with the fact that you exert an attractive force on the Earth as well as the Earth exerting an attractive force on you.

5.9 Momentum is conserved

Newton's third law of motion leads on to an important new law about collisions. Imagine two trolleys, for example, running towards each other, colliding and then rebounding from each other, as in Fig. 5.12. While they are touching, whatever force trolley A exerts on trolley B is the same as the force that B exerts on A, but in the opposite direction.

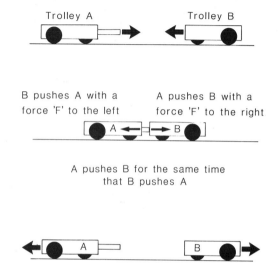

Fig. 5.12

Trolley A obviously pushes trolley B for exactly the same time as B pushes A, so the impulse (force×time) on A is equal but opposite to the impulse on B. An impulse results in a change in momentum, so A and B change their momenta by equal but opposite amounts.

What do we mean by *opposite* amounts? Momentum is a vector quantity; it has a direction as well as a magnitude. You must remember to take the direction into account. If you imagine one direction is *positive* then the opposite direction will be *negative*. So A gains *positive* momentum while B gains *negative* momentum; in other words, B *loses* momentum. B loses the same quantity of momentum that A gains, so the *total quantity* of momentum stays the same (effectively, in this case, B has given A some of its momentum).

This is an example of the **law of conservation of momentum**, which states:

Whenever two or more objects exert forces on each other, (such as in a collision), the total momentum remains constant.

Notice that there does not have to be a collision for this law to apply. Think again of falling off a chair. You gain momentum (downwards). Remember that the Earth is pulled up with an equal force to the force on you, so it gains momentum upwards. The momentum of you and the momentum of the Earth are equal but opposite, so the total momentum remains constant—zero in this case, since you have zero momentum as you start to fall off the chair.

There is no collision in this example, until, of course, you hit the floor! You then lose all your momentum. What happens to the Earth's momentum?

'Conservation laws' are important to physicists. It helps to make sense of the world around us to be able to find things which stay the same all the time, that is, which are conserved. There are not many such things. Momentum is one; two others are energy and electric charge.

5.10 Testing the law of conservation of momentum

You should be able to design for yourself some experiments to test whether the law of conservation of momentum applies during collisions. The illustrations in this section should help to give you some ideas.

Figs. 5.13 to 5.16 suggest possible ways of using colliding trolleys to investigate momentum. In each of your experiments you will need to measure the total momentum of the trolleys before the collision, and compare it to the total momentum of the trolleys after the collision. The law of conservation of momentum says that the two values should be equal.

Fig. 5.17 suggests how you might take a stroboscopic photograph of two colliding air-track vehicles in order to measure their velocities and hence calculate the momenta. The vehicles can be identified on the photographs by sticking a long straw in one vehicle and a short straw in the other.

You should also be able to think of ways of using other measuring instruments that you have met to investigate collisions of air-track vehicles.

Figs 5.13–5.16   Suggestions for investigating the law of
conservation of momentum

Fig. 5.13

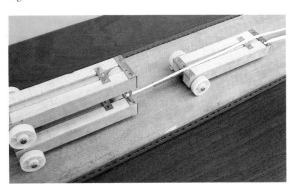

Fig. 5.14

Fig. 5.15

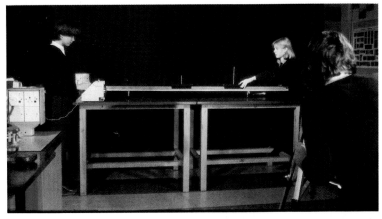

Fig. 5.16

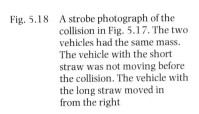

Fig. 5.17   Taking a strobe photograph of
a collision on an air track

Fig. 5.18   A strobe photograph of the
collision in Fig. 5.17. The two
vehicles had the same mass.
The vehicle with the short
straw was not moving before
the collision. The vehicle with
the long straw moved in
from the right

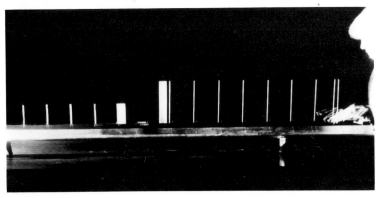

5.11 A numerical example using the law of conservation of momentum

A railway wagon of mass 10 tonnes is shunted with a velocity of 2 m/s into a siding where it bumps into four identical stationary wagons; the wagons couple together and move off together after the collision. With what velocity do all five wagons move after the collision?

Before collision:

mass of moving wagon $m_1 = 10\,000\,\text{kg}$
velocity of moving wagon $v_1 = 2\,\text{m/s}$

$$\begin{aligned}
\text{momentum of moving wagon} &= m_1 \times v_1 \\
&= 10\,000\,\text{kg} \times 2\,\text{m/s} \\
&= 20\,000\,\text{kg\,m/s}
\end{aligned}$$

Since the four coupled wagons are not moving before the collision, they have no momentum. Hence the total momentum before the collision is 20 000 kg m/s.

After collision:

mass of moving wagons $m_2 = 50\,000\,\text{kg}$
velocity of moving wagons $v_2 = ?$

$$\begin{aligned}
\text{momentum of moving wagons} &= m_2 \times v_2 \\
&= 50\,000\,\text{kg} \times v_2
\end{aligned}$$

momentum before collision

$$= \text{momentum after collision}$$
$$20\,000\,\text{kg\,m/s} = 50\,000\,\text{kg} \times v_2$$
$$v_2 = \frac{20\,000\,\text{kg\,m/s}}{50\,000\,\text{kg}}$$
$$= 0.4\,\text{m/s}$$

The trucks move off with a velocity of 0.4 m/s.

You may be able to see that there is a quicker way of solving this problem. When the single truck collides with the other four, the total mass of the trucks which are moving becomes *five times larger* than it was before (the mass increases from 10 tonnes to 50 tonnes). For momentum to be conserved, the velocity of the moving trucks must become *five times smaller* than it was before, that is, it must decrease to $\frac{1}{5}$ of 2 m/s, which is 0·4 m/s.

5.12 Momentum conservation in action

Why does a gun recoil when it is fired? How does a rocket motor or a jet engine push something along? How does a ship's propeller move a ship? If you

Fig. 5.19   Mons Meg (at Edinburgh Castle)

jump from a rowing boat onto a river bank, why are you in danger of getting wet? All these questions can be answered using the idea of momentum.

Imagine the gun in Fig. 5.19 firing one of those cannon balls. Before the gun is fired, neither the gun nor the cannon ball has any momentum; the total momentum is zero. After the gun is fired, the cannon ball will have momentum in one direction, so the gun must have an *equal* momentum in the opposite direction so that the *total* momentum remains zero; the gun 'recoils'. It was not a good idea to stand just behind such a gun when it was being fired.

Each of the cannon balls has a mass of 50 kg. If one were fired at a velocity of 80 m/s, how much momentum would it have? How much momentum would the gun have? If the mass of the gun is 2000 kg, with what velocity would it recoil?

You could use this idea to move the gun along a road. When it recoils it soon stops moving again because friction provides a stopping force. Another cannon ball could then be fired to move the gun a little further. This is a rather noisy, destructive and energy-wasting method of moving a gun, but rocket and jet engines work using this idea. They do not

Fig. 5.20   A jet engine

throw out cannon balls; they throw out millions of gas molecules every second at a very high speed. Think of them as very tiny cannon balls! The gas molecules gain momentum in one direction, so the rocket or jet gains an equal quantity of momentum in the opposite direction.

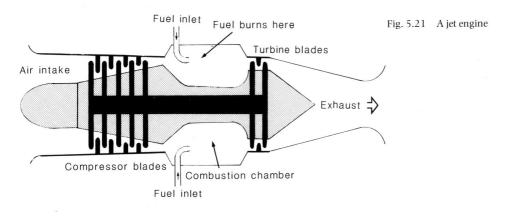

Fig. 5.21   A jet engine

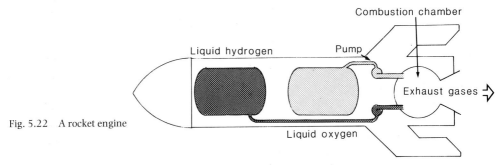

Fig. 5.22   A rocket engine

The molecules that a jet engine throws out are a mixture of air molecules that have been sucked into the engine and fuel molecules that have burned in the oxygen from the air. A rocket engine carries its own supply of oxygen in one form or another with which to burn its fuel, so a rocket engine can go where there is no air, whereas a jet engine cannot.

A ship's propeller gives water momentum in one direction, so the ship gains an equal momentum in the opposite direction. Why does the ship not gain more and more momentum and move faster and faster?

Using the ideas in this section, try to explain why there is a danger of getting wet if you jump from a rowing boat to a river bank.

Fig. 5.23   A ship's propeller

SUMMARY

Now that you have finished studying this chapter on collisions, there are a number of things you should know or be able to do.

1 You should:
   a) realize that things with a large momentum can cause more damage than things with a small momentum;
   b) understand that, the longer something takes to stop, the smaller the stopping force required;
   c) know that momentum is calculated by multiplying mass by velocity;
   d) know that impulse is calculated by multiplying force by time;
   e) know that an impulse gives rise to a change in momentum, and be able to perform simple numerical calculations using this fact;
   f) understand Newton's third law of motion;
   g) understand the law of conservation of momentum;
   h) be able to perform and describe an experiment to test the law of conservation of momentum;
   i) be able to solve simple numerical problems involving conservation of momentum;
   j) understand why a gun recoils;
   k) understand how rocket or jet engines, and propellers, can cause something to move.

2 You should know what each of the following is, or what each does:

attractive force	repulsive force
molecule	rocket engine
jet engine	reaction

3 Here are some of the other words that have been used in this chapter. You should know what each word means:

sub-atomic	represent
numerical	gradual
negligible	conserve
interpret	identical
destructive	consider
combination	exert
stationary	

FURTHER QUESTIONS

1 Peter and Paul are each on a pair of roller skates, as shown in Fig. 5.24. Peter has twice the mass of Paul. Peter pulls Paul towards him with a rope, as shown.

Fig. 5.24   Peter   Mass = 2m   Paul   Mass = m

   a) Peter will also start moving towards Paul. Explain why.
   b) Will both boys move with the same acceleration? If not, who has the greater acceleration and why?
   c) Exactly where will the two boys bump into each other?

2 A rifle, of mass 2 kg, fires a bullet of mass 0.1 kg at a velocity of 50 m/s.
   a) Explain why the gun moves back, or 'recoils'.
   b) With what velocity does the gun recoil?
   c) Explain why somebody firing a rifle braces the butt of the rifle against their shoulder.

3 Here are details of three moving objects:
   (i) A charging rhinoceros of mass 2500 kg moving at 10 m/s.
   (ii) A bicycle and rider of mass 80 kg moving at 10 m/s.
   (iii) A lorry of mass 2500 kg moving at 30 m/s.
   a) Which of these has the greatest momentum? What is the value of that momentum?
   b) Which has the least momentum? What is the value of that momentum?
   c) Which object would be the hardest to stop?

4 Jill jumps from a diving board into a swimming pool full of water. Jenny jumps from the same board onto the side of the pool. Who is more likely to get hurt? Explain your answer using the ideas you have learned in this chapter.

5 Fig. 5.6 showed Toby landing on the ground after jumping from a wall. Toby has a mass of 50 kg. If he landed with a speed of 5 m/s, what was the force making him decelerate if:
   a) he bent his legs and took 1 s to stop;
   b) he forgot to bend his legs and stopped in 0.01 s?

6

CIRCULAR MOTION

Fig. 6.1

6.1 Introduction

Has it ever occurred to you how much physics there is to be found in a fairground? You have already seen bumper cars at the start of the last chapter; now here is another example—the 'big wheel', an example of motion in a *circle*.

Can you think of any other examples of physics to be found in a fairground?

Ever since the wheel was invented, man has depended on circular motion for nearly all forms of transport. Look back at the photographs at the start of Chapter 2. Which pictures include something which is moving in a circle? In which object do you think you would find the greatest number of things moving in circles?

In this chapter you will see how the ideas that you have already met from looking at motion in a straight line can be extended to motion in a circle.

6.2 You should know . . .

In order to understand this chapter, you should have learned the physics of motion in a straight line in Chapters 2, 3 and 4 in this book. In particular, you should remember:
1 that an unbalanced force causes an acceleration;
2 that velocity is a vector quantity, that is, it has both a size (magnitude) and direction.

You also need to know a little about circles, such as what the circumference of a circle is and how to calculate the circumference if you know the radius or the diameter.

6.3   Centripetal force

Fig. 6.2   What forces are acting on this hammer?

Look at the competitor in the highland games in Fig. 6.2 whirling a hammer over his head. Even if the hammer is whirled at the same speed all the time, does it travel at a constant *velocity*? Remember that velocity is a vector quantity, and has both size and direction. Is the hammer going in the same direction all the time? Obviously not! Its direction is changing so its velocity is changing; that is to say, it is *accelerating*. This may seem a strange statement to you. You usually think of acceleration as meaning 'going faster', but to a physicist an acceleration is a *change of velocity*, so a change of direction, even without a change of speed, is still an acceleration.

Remember that Newton's first law of motion says that, if the hammer is accelerating, there must be a *force* acting on it to make it accelerate. The rope provides the force on the hammer, pulling it. (It cannot *push* it—why not?) The direction of the force must be towards the *centre of the circle* as shown in Fig. 6.3. Notice that this force is acting perpendicular to the direction in which the hammer is moving.

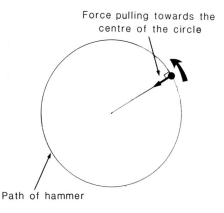

Force pulling towards the centre of the circle

Path of hammer

Fig. 6.3   Centripetal force

A force like this which acts towards the centre of a circle and makes something follow a circular path is called a **centripetal force**, and it causes a **centripetal acceleration**.

What happens when the competitor lets go of the rope? There will then be no centripetal force acting on the hammer, so it will stop going in a circle. According to Newton's first law, if there are no forces acting on the hammer it will keep going in a straight line. It will go in whatever direction it was moving when the competitor let go of the rope. That direction is at a tangent to the circle, as shown in Fig. 6.4.

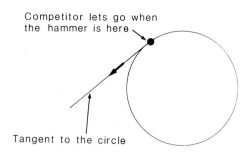

Competitor lets go when the hammer is here

Tangent to the circle

Fig. 6.4

Although the hammer moved off at a tangent to the circle when the competitor let go of the rope, it did not actually keep going in a straight line. Why not? What happened to it?

6.4   Cars and corners

Fig. 6.5   What forces push this car round the corner?

You have just learned that a centripetal force is needed to make something go round in a circle.

What actually provides the force to make things go round in a circle? Let us look at a number of examples.

Fig. 6.5 shows a common example. Obviously the friction between the tyres and the road provides the force which pushes the car round the corner. If there is not enough friction (for example, if the road is icy) what happens to a car trying to go round a corner?

If you were a passenger in this car, you may feel as if you were being 'flung outwards' as the car goes round the corner. You may have heard of something called 'centrifugal force' flinging you outwards. As far as we are concerned in this book, there is *no such thing as centrifugal force*. We can explain 'being flung out' as follows.

If you are sitting in the car, you must go round the corner with the car, so you must have a centripetal force acting on you to pull you round the corner. What provides this centripetal force? Fig. 6.6 shows two possibilities; it could be friction between you and the seat, or it could be the side of the car pushing you round.

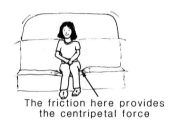

The friction here provides
the centripetal force

The side of the car provides
a centripetal force

Fig. 6.6   How a centripetal force can act on passengers in a car

Imagine you are sitting in the middle of the rear seat. Suppose there is *not enough* friction to provide the centripetal force needed to pull you round the circle with the car. You will tend to move on in your original direction (not quite in a straight line—there is bound to be *some* friction) while the car 'slides away under you' (Fig. 6.7). This sliding will stop when one side of the car hits you and pushes you round in the same circle as the car. You *feel* as if you have been 'flung out' to the side of the car.

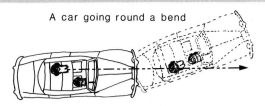

A car going round a bend

The passenger continues to move
in a straight line if there is
no centripetal force

Fig. 6.7   What happens if there is not enough centripetal force

It is not only road vehicles which go round corners. Look at Fig. 6.8. This is a photograph of a pair of railway carriage wheels. Notice the flanges on the inside of the wheels. How do you think the flanges give the carriage a centripetal force to make it go round a curve? Is it the flanges on the wheels on the outside or inside of the curve which are responsible for the centripetal force?

Fig. 6.8   Railway carriage wheels. Note the flanges

6.5   Spin dryers

Fig. 6.9   A spin dryer

You should now understand sufficient about circular motion to be able to understand how a spin dryer can remove most of the water from wet clothes. As the spin dryer goes round, the clothes go round in a circle with it. What provides the centripetal force on the clothes?

The fibres of the clothing cannot provide sufficient centripetal force on most of the water to make it go round in a circle, so it tends to go at a tangent to the circle, until it reaches the side of the tub. What do you think happens to it there? (Look at Fig. 6.9.)

The water does not actually go at a *tangent* to the circle; the clothes do exert *some* force on the water. Fig. 6.10 shows the path some water might take to reach the side of the drum.

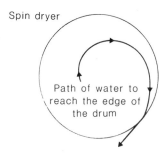

Fig. 6.10   The path of some water in a spin dryer

6.6   Satellites

In this section you are going to use your knowledge of motion, particularly circular motion, to understand why satellites can go round the Earth. You will use some of the ideas you met in Section 4.8 'Falling sideways'.

Imagine a high mountain with a large gun at the top of it. The gun is fired. Gravity pulls the shell down so it flies in a parabolic path to hit the ground some distance away (as shown in Fig. 6.11) in the manner you learned in Section 4.8.

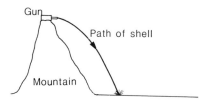

Fig. 6.11   Firing a gun from a mountain top

Suppose the shell is fired so that it flies faster. It will obviously travel further from the gun [Fig. 6.12(a)]. But this diagram is not accurate—the Earth is curved, so the shell actually travels a little further, as shown in Fig. 6.12(b). The faster the shell is fired, the further it travels before it hits the

Earth, until there comes a speed at which it never hits the Earth—the Earth is curving away underneath it at exactly the same rate at which the shell is falling. The shell orbits the Earth; it has become a **satellite**. Its weight provides the centripetal force needed to make it move in a circle.

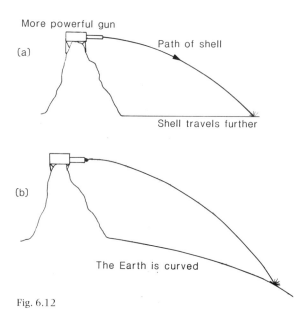

Fig. 6.12

The speed of the satellite must be just right for this centripetal force; too slow and it falls to the ground (Fig. 6.12); too fast and it flies out into space (Fig. 6.14). What is the right speed?

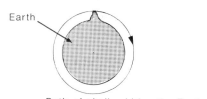

Fig. 6.13

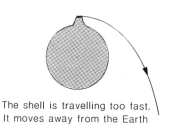

Fig. 6.14

You can solve this problem with the aid of a *scale diagram* of the Earth and the satellite orbit. Fig. 6.15 shows you how. Use a scale of 1 mm to represent 2 km. The radius of the Earth is about 6400 km, which is 3.2 m on this scale.

(a) Draw a curve of radius 3.2 m to represent part of the Earth's surface

Earth

(b) Draw a second curve 75 mm above the first to represent part of the path of the satellite

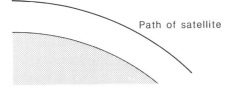

Path of satellite

(c) Draw a tangent to the satellite's path

The satellite would follow this path if there were <u>no centripetal force pulling it into a circle</u>

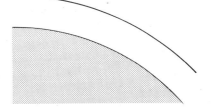

(d) If the satellite travels this far in 100 s, while it is also falling 50 km, it will stay on its proper orbit

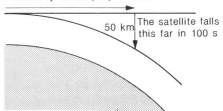

50 km | The satellite falls this far in 100 s

Fig. 6.15   Calculating the orbital speed of a satellite

Imagine your satellite is 150 km above the Earth (75 mm on this scale). Using a suitable length of string and a pencil, draw on a sheet of paper an arc of radius 3.2 m to represent the Earth (there is no need to draw the whole Earth—an arc about 80 cm in length is sufficient) and a second arc 75 mm above the first to represent the satellite's orbit. You must work out a sensible way of doing this, but Fig. 6.16 might help.

Suppose the satellite was fired horizontally. If it did not fall towards the Earth, it would keep going in a straight line with constant velocity [Fig. 6.15(c)]. However, it does fall towards the Earth with an acceleration of $10 \, \text{m/s}^2$, which is why it goes in a curve.

Let us think of what happens in 100 seconds. We can calculate how *far* the satellite falls in this time by using

$$s = ut + \tfrac{1}{2}at^2$$
$$s = 0 + \tfrac{1}{2} \times 10 \, \text{m/s}^2 \times 100 \, \text{s} \times 100 \, \text{s}$$
$$= 50\,000 \, \text{m} \quad \text{(or 50 km)}$$

This is 25 mm on your scale diagram.

We must give the satellite sufficient velocity so that, after 100 s, it would be 50 km above its orbit if the Earth did not make it fall. Since it *does* fall 50 km, it will 'fall' onto its orbit. If the satellite moves too slowly, it will fall below its orbit; if too fast it will fall above its orbit.

Fig. 6.16   Drawing a scale diagram

Name	Launch date	Actual or estimated lifetime	Period (min)	Average height (km)	Notes
Sputnik 1	4 Oct 1957	3 months	96.2	580	First artificial satellite
Explorer 1	1 Feb 1958	10 years	114.8	1450	First US satellite
Tiros 1	1 April 1960	60 years	99.2	725	First weather satellite
Echo 1	12 Aug 1960	7 years	118.2	1600	First passive communications satellite
Vostock 1	12 April 1961	1 orbit	89.3	240	First manned space flight
Telstar 1	10 July 1962	10 000 years	157.7	3290	First active communications satellite
Early Bird	6 April 1965	10^6 years	1437	36000	First commercial synchronous communications satellite
Asterix 1	26 Nov 1965	200 years	108.6	1170	First French satellite

Find the point on your diagram where the satellite will be 25 mm above its orbit if it does not fall [Fig. 6.15(d)]. The distance from the start to this point must be how far the satellite travels in 100 s. Measure this distance on your diagram.

How many kilometres is this distance when scaled up?

What is the satellite's speed (in km/s)?

What distance does the satellite travel in going once round the Earth?

How long will it take to travel round the Earth?

The first artificial earth satellite, Sputnik 1, was launched on 4th October 1957 by the Soviet Union. It had an elliptical orbit rather than a circular one. The average height was about 580 km and it took 96.2 minutes to orbit the Earth. The table above gives some details of other early satellites.

Does the period you have calculated for a satellite roughly agree with that for Sputnik 1, or any other satellite? Why do you think Early Bird has a much longer period than any other satellite listed?

6.7 Weightlessness

People often say that an astronaut in an orbiting spacecraft is weightless. They are wrong! The Earth's gravitational field is still acting on the astronaut; after all, gravity is responsible for the centripetal force on the spacecraft. It is true that the astronaut *feels* weightless, but this is not quite the same thing.

What gives *you* a feeling of weight? The floor, or chair, which is supporting you exerts an upwards force on you; you can *feel* this force between you and the floor or chair. If you jump into a swimming pool, for example, for a short time there is nothing to support you and so you lose any sensation of weight. Of course, you still *have* a weight—it is accelerating you downwards, but you cannot feel it.

An orbiting astronaut is also falling, at the same rate as the spacecraft, so the spacecraft is not holding the astronaut up, so there is no sensation of weight. The astronaut's weight is actually providing the centripetal force needed to make him or her move in a circle.

6.8 How large is a centripetal force?

Before you can go much further in understanding circular motion, you need to be able to calculate centripetal forces so that you can make predictions about bodies moving in circular orbits.

The derivation below makes use of the 'crossed chord theorem' from geometry. You do not have to remember this derivation, but do try to follow and understand it if you can.

The circle in Fig. 6.17 represents the orbit of a body of mass m moving with a speed v. Suppose that in a time t, it travels from A to B. It has 'fallen' a distance h towards the centre of the circle, as well as moving a distance x sideways. The acceleration 'downwards' (towards the centre) is related to the distance h by the equation $s = ut + \frac{1}{2}at^2$. In this case:

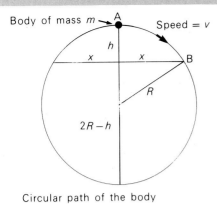

Fig. 6.17

initial velocity $\quad u = 0 \quad$ (Particle not moving down at A.)

$\qquad$ distance $\quad s = h$

hence $\quad h = \frac{1}{2}at^2$

or $\qquad a = \dfrac{2h}{t^2} \quad$ (equation 1)

This is the acceleration 'downwards' towards the centre of the circle, i.e. the centripetal acceleration. This equation is of no use as it stands, with h and t in it. Why not? We need an equation with constant values in it, like R and v.

According to the crossed chord theorem (and referring to Fig. 6.17):

$$h(2R - h) = x^2$$

so $\qquad h = \dfrac{x^2}{2R - h}$

Since $h \ll R,$ $\qquad h = \dfrac{x^2}{2R}$

If x is very small, $x =$ arc length AB.

The particle travels from A to B in time t, so

$$x = vt$$

therefore $\qquad h = \dfrac{(vt)^2}{2R}$

Substitute this into equation 1 above:

$$a = \frac{2(vt)^2}{t^2 2R}$$

$$= \frac{v^2}{R}$$

If x is very small indeed, this is the 'instantaneous' centripetal acceleration at A.

Since $F = ma$, the centripetal force on a body is given by

$$F = \frac{mv^2}{R}$$

Does this equation agree with common sense? The equation predicts that the force F will increase if either the mass m or the speed v increases, or if the radius r of the circle decreases. Does this seem reasonable?

6.9 Testing $F = mv^2/r$

The equation we have derived above is a *theoretical prediction*; the next task is to carry out an experiment to *test* this prediction.

Figs 6.18 and 6.19 show one possible way of doing this. The centripetal force to keep the bung moving in a circle is provided by the weight hanging on the end of the string. You will find a weight of 1 N (the weight of a 100 g mass) is suitable.

Fig. 6.18 Using the centripetal force kit

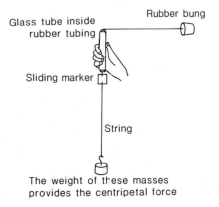

Fig. 6.19

To use this apparatus, adjust the sliding marker so that, when the indicator is just touching the bottom of the glass tube, there is 0.5 m of string coming out of the top of the tube. This will be the radius of the bung's circular orbit. Hold the glass tube, stand well clear of other people and any furniture, and whirl the bung round your head in a horizontal circle at the right speed for the indicator to be *just* touching the bottom of the tube. You will probably need to practise this for a time before you are confident you have the right speed.

Your partner should record the time taken for ten revolutions of the bung, using a stopwatch.

Now answer these questions.

1 How long did it take for your bung to do ten revolutions?
2 How long did it take to do 1 revolution?
3 What was the radius r of the bung's orbit?
4 What was the circumference of the orbit (i.e. the distance the bung travelled in 1 revolution)?
5 What, then, was the speed of the bung?
6 What is the mass m of your bung (in kg)?
7 Now calculate mv^2/r. Is it roughly equal to the weight hanging on the string providing the centripetal force?

Repeat the experiment a few more times using different radii or weights. Do your experiments suggest that the equation mv^2/r is the correct way of calculating the centripetal force?

Do you think this is a reliable experiment? What are the main problems?

6.10 Satellites and the Moon

Now that you know that centripetal acceleration can be calculated from the equation $F = mv^2/r$, you can *calculate* the time taken for a satellite to orbit the Earth, instead of drawing a scale diagram in the way you did in Fig. 6.16.

In the case of a satellite the centripetal acceleration is the acceleration of free-fall, i.e. $10\,\text{m/s}^2$; the radius of the orbit of the satellite in Fig. 6.15 was 6550 km (6 550 000 m).

$$\text{acceleration} = \frac{v^2}{r}$$

$$10\,\text{m/s}^2 = \frac{v^2}{6\,550\,000\,\text{m}}$$

$$v^2 = 65\,500\,000\,(\text{m/s})^2$$
$$v = 8090\,\text{m/s}$$

How long will it take to complete an orbit at this speed? Does this agree with the value you obtained from your scale diagram?

Notice that you do not need to know the mass of the satellite; the acceleration is the same for all objects in the gravitational field, and hence the time of orbit is the same for all objects.

The moon is a satellite of the Earth, so we can do the same calculation to find the time of orbit of the Moon. The radius of the Moon's orbit is 384 000 km ($= 3.84 \times 10^8$ m) so

$$\text{acceleration} = \frac{v^2}{r}$$

$$10\,\text{m/s}^2 = \frac{v^2}{3.84 \times 10^8\,\text{m}}$$

$$v = \sqrt{3.84 \times 10^9}\,\text{m/s}$$
$$= 6.20 \times 10^4\,\text{m/s}$$

The distance round one orbit is

$$2\pi r = 2 \times 3.14 \times 3.84 \times 10^8\,\text{m}$$
$$= 2.41 \times 10^9\,\text{m}$$

Therefore the time taken to go round one orbit is

$$\frac{2.41 \times 10^9\,\text{m}}{6.20 \times 10^4\,\text{m/s}}$$

$$= 3.89 \times 10^4\,\text{s}$$

This is nearly 11 hours.

Oh dear! Something has gone very wrong! The Moon actually takes about 27 days to orbit the Earth and we have calculated a time of 11 hours. Where is the mistake?

In 1665 Isaac Newton, who was studying at Trinity College, Cambridge, returned to his home in Woolsthorpe, in Lincolnshire, to escape the plague. He remained there for two years and in that time produced many of his great ideas. These include the laws of motion and the idea of gravitation. The story goes that it was the sight of an apple falling to the ground, pulled by gravity, that gave him the idea that the Earth's gravitational field supplied the centripetal force on the Moon to keep it in its orbit (a fact that has been taken for granted in this book so far). He reasoned that the Earth's gravitational field must be much *less* than 10 N/kg at the distance of the Moon's orbit to account for the long time taken for the Moon to complete one orbit.

The simplest suggestion is that the gravitational field obeys an inverse proportionality law, that is, the field halves as the distance from the centre of the Earth doubles. The Moon is 60 times further from the centre of the Earth than you are, so, if gravitational fields obeyed this law, the field at the Moon would be $\frac{1}{60} \times 10 \, \text{N/kg} = \frac{1}{6} \, \text{N/kg}$.

If you repeat the calculation above, the speed of the Moon turns out to be 8000 m/s, giving a time of orbit of about $3\frac{1}{2}$ days. (Check these calculations for yourself.)

Wrong again!

Faced with a similar result, Newton wondered if gravity obeyed an *inverse square law*, that is, as the distance from the Earth's centre is doubled, the gravitational field falls to a quarter of its original value; if the distance is trebled, the field becomes $\frac{1}{9}$ of its original value, and so on.

If this were so, the gravitational field due to the Earth felt by the Moon would be

$$\frac{1}{60} \times \frac{1}{60} \times 10 \, \text{N/kg} = \frac{1}{360} \, \text{N/kg}$$

Repeating the above calculation yet again gives the speed of the Moon as 1033 m/s, which leads to an orbit time of about 27 days, which, of course, fits with observations.

Newton went on to suggest that *all* bodies attract each other with a gravitational force which depends on the masses of the bodies concerned and which obeys an inverse square law for distance. It is this force which keeps the Earth and other planets in orbit round the Sun, and keeps galaxies of stars spinning like vast Catherine wheels.

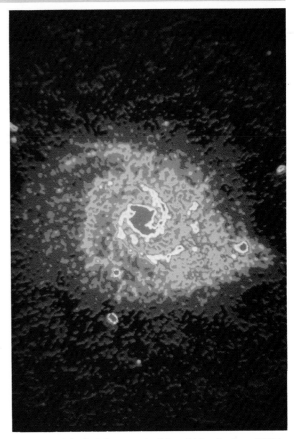

Fig. 6.20   A 'radio' photograph of the whirlpool galaxy (M51)

SUMMARY

Now that you have finished studying this chapter on circular motion, there are a number of things you should know and be able to do.

1 You should:
   a) know that anything moving in a circle has a centripetal force acting on it;
   b) know that a centripetal force acts towards the centre of the circle;
   c) be able to explain why you tend to be flung to the outside of a car going round a corner;
   d) appreciate the importance of friction between car tyres and the road for cars going round a corner;
   e) be able to explain how a spin dryer dries clothes;
   f) be able to explain why a satellite remains in orbit;
   g) know why an astronaut feels weightless;
   h) know how to use the equation 'centripetal force = mv^2/R';
   i) know how to carry out one experiment to test the centripetal force equation;
   j) know that the Earth's gravitational field obeys an inverse square law.

2 You should know what each of the following is, or what each does:

tangent (to circle)	revolution
centripetal	satellite
acceleration	astronaut
flange	inverse square law
orbit	

3 These are some of the other words that have been used in this chapter. You should know what each word means:

whirl	artificial
necessary	responsible
sensation	prediction

FURTHER QUESTIONS

1 The following are all examples of things moving in a circle. Write down what provides the centripetal force in each case.
   a) The Moon moving round the Earth.
   b) A penny on the rotating turntable of a record player.
   c) A conker being whirled round on the end of a piece of string.
   d) A car going round a corner.
   e) A train moving round a curve.
   f) Clothes inside a rotating spin dryer.
   g) Passengers sitting on a seat in a car moving round a corner.

2 Fig. 6.21 shows the path of a stone tied to a length of string which is held at 0. The stone is being swung round in a vertical circle.

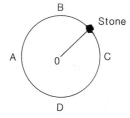

Fig. 6.21

   a) What provides the centripetal force to keep the stone moving in a circle when it is at position A or C?
   b) What *two* forces combine to provide the centripetal force on the stone when it is at B?

   c) If the string is weak it is most likely to break at position D. Why is this?
   d) Suppose the string does break when the stone is at position D. Draw a diagram to show the path of the stone after the string breaks.

3 A ball of mass 200 grams is attached to one end of a string and whirled round in a horizontal circle of radius 1 metre once every second.
   a) What is the speed of the ball (in metres per second)?
   b) What centripetal force does the string exert on the ball?
   c) Repeat parts a) and b) for a circle of radius 0.5 m.
   d) From the equation mv^2/r it might appear that the smaller the radius of the circle, the larger will be the centripetal force (since you have to *divide* by r). Your answers to b) and c) should show that this is not so. Explain why.

4 Some fairgrounds have a large 'Rotor', which is a large upright drum inside which people stand with their backs against the wall. When the drum is rotated at full speed, the floor is pulled down away from the occupants, who are left apparently glued against the wall. Explain why the occupants do not go down with the floor.

5 It is possible to fill a bucket with water and then whirl it round in a vertical circle without spilling any water, even when the bucket is upside down. Explain how this is possible.

7

MEASURING ENERGY

Fig. 7.1

7.1   Introduction

You already know (perhaps from reading Chapter 1) that the gravitational potential energy of this falling water is being converted to kinetic energy.

How much gravitational potential energy is the water converting to kinetic energy every second? If you wanted to use the energy of the falling water to generate electrical energy, you would have to know how to answer this question. Engineers who design hydroelectric power stations (such as those in Figs 1.7 and 1.8) must make sure that the energy that is obtained from the falling water is sufficient for the size of the power station that is being built.

In this chapter you will begin to learn how to measure energy. As you know, there are many different kinds of energy; a single chapter cannot contain the details of how to measure all the different kinds. To start with, you will learn about gravitational potential energy, elastic potential energy and kinetic energy.

7.2   You should know . . .

In order to understand this chapter, you should:
1 know about and be able to recognize different kinds of energy:
2 have read Chapter 1 of this book;
3 know the definition of 1 newton of force.

61

7.3 Measuring gravitational potential energy

Fig. 7.2

Fig. 7.3

Fig. 7.4

Suppose you lift a book from the floor to a table, like Toby in the photograph. You will have to transfer some energy from your store of chemical energy in your muscles to the book. The book gains gravitational potential energy.

If you lift *two* identical books, you will transfer *twice* as much energy, three books, three times as much energy, and so on.

Suppose, instead of lifting the book onto the table, you lift it on top of two tables piled on top of each other, so you have to lift the book twice as far from the ground as before. How much more energy do you need to lift a book to the height of two tables than to the height of one table?

When you lift the book, the energy you transfer depends on two things:

1 the *weight* of the book—the greater the weight of book the bigger the force you have to exert and the more energy is transferred to gravitational potential energy of the book;
2 the *distance you lift* the book—the further you lift the book, the more energy is transferred to gravitational potential energy.

Physicists have agreed that a sensible way of measuring the transfer of energy would be to multiply the force by the distance:

energy transferred
= force × distance moved by force

Another phrase which means the same thing as 'energy transferred' is 'work done', so we could also write this definition as

work done = force × distance moved by force

Our unit of force is a *newton*, and our unit of distance is a *metre*, so a unit of energy could be a **newton metre**. A unit of energy *is* sometimes written like this, but a newton metre is usually called a **joule** of energy. The definition of a joule of energy is:

1 joule of energy is transferred when a force of 1 newton moves for a distance of 1 metre.

How much energy was needed to lift a book onto the table? The mass of the book Toby was lifting in Fig. 7.2 was 0.5 kg.

The weight of the book is $m \times g$
$$= 0.5 \text{ kg} \times 10 \text{ N/kg}$$
$$= 5 \text{ N}$$

This is the force needed to lift the book.

The height of the table was 0.6 m. Therefore:

energy transferred = force × distance
$$= 5 \text{ N} \times 0.6 \text{ m}$$
$$= 3 \text{ joules} \quad \text{(or 3 J)}$$

(Notice we use a capital 'J' as the abbreviation for joule.)

The amount of energy Toby transferred to the book was 3 joules and so this is the amount of gravitational potential energy gained by the book. You can see that you can calculate the gravitational potential energy of anything by multiplying its *weight* by its *height above the floor*. Since weight is mass × gravitational field strength (g), we can write

gravitational p.e. = $m \times g \times h$

where m = mass, g = gravitational field strength and h = height above floor.

You will usually see the equation for calculating gravitational potential energy written in this way.

To give you some practice using this idea of measuring gravitational potential energy, calculate the following (you will need to measure some heights and some weights, or masses and multiply by g).

How much gravitational potential energy:

1 does this book gain when it is lifted from the floor to your desk or bench;
2 does your school-bag gain when you lift it off the floor to carry it;
3 do you gain when you walk up the nearest flight of stairs or steps?

7.4 Elastic potential energy

Fig. 7.5   What energy changes occur here?

How much energy is stored in the catapult in Fig. 7.5? Richard has had to pull back the rubber, so he has transferred energy from his muscles to the rubber. This energy is now stored in the catapult as 'elastic potential energy'. To calculate the energy stored, it seems that you need to measure the force needed to pull back the rubber and the distance it has been pulled, as in Fig. 7.6, and then multiply the two together as in the last section.

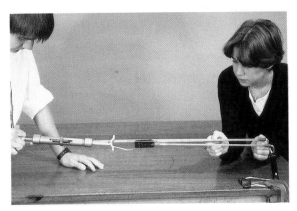

Fig. 7.6   How do you calculate elastic potential energy from these measurements?

But wait a minute! You do not pull the rubber back with a *constant* force. It is easy to pull to start with, and becomes more and more difficult as the rubber is stretched further. Would it not be better if you used the *average* force needed to stretch the rubber to calculate the energy stored? If, when you start to pull the rubber, you need virtually no force, is not the average force going to be the force in the 'middle' of the pull, that is, half the final force that you see on the forcemeter in Fig. 7.6?

The final force you need to keep the rubber stretched is called the 'tension' in the rubber. So the way to calculate the energy stored in the rubber is by using

$$\text{elastic potential energy} = \tfrac{1}{2} \times \text{tension} \times \text{extension}$$

7.5 The area under a force-extension graph

You have probably investigated the way in which a spring stretches when a load is hung on the end. A graph of force against extension is usually a straight line, at least to start with, like the one in Fig. 7.7. A spring whose force-extension graph is like this, so that the force is proportional to the extension, is said to obey **Hooke's law.**

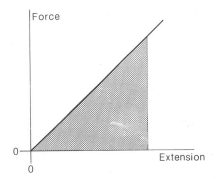

Fig. 7.7   A force-extension graph for a spring

Look at the area under the graph which has been shaded in Fig. 7.7. This is a triangle and its area is

$$\tfrac{1}{2} \times \text{base} \times \text{height}$$
$$= \tfrac{1}{2} \times \text{extension} \times \text{tension in spring}$$

In the last section, we did exactly the same calculation in order to work out the energy stored in the catapult. The area under a force-extension graph is the energy stored in the spring or rubber. As with the velocity-time graphs in Chapter 2, this is true whether or not the graph is a straight line. Of course, if the graph is not a straight line, you will have to find the area not of a simple triangle, but of a more complicated shape.

You can now see that the equation we used in the last section:

$$\text{energy stored} = \tfrac{1}{2} \times \text{tension} \times \text{extension}$$

only applies if the force-extension graph is a straight line, i.e. the spring or rubber obeys Hooke's law. If the graph is not a straight line the average tension will not be half the final tension.

Were we right to use this equation in the last section? Does rubber obey Hooke's law? Design and carry out an experiment to find out.

7.6 Kinetic energy

After the stone had left the catapult in Fig. 7.5 it had kinetic energy. How much kinetic energy? We usually assume that all the elastic potential energy stored in the rubber became kinetic energy of the stone. If you know how much elastic energy there was, you know how much kinetic energy the stone has.

But you do not always know where a moving object gets its energy from, or you cannot calculate it even if you do know. This section shows you how to calculate the kinetic energy of a moving object from its mass and velocity.

Imagine a frictionless vehicle of mass m as illustrated in Fig. 7.8. Suppose you pushed it with a force F for a distance s. You would transfer to this vehicle an amount of energy equal to $F \times s$. Since there is no friction all this energy would become kinetic energy.

We can summarize this as follows:
'The vehicle gains kinetic energy as a result of the work done ($F \times s$).'

This can be written in mathematical language as

$$\text{k.e.} = F \times s$$

The force F causes the vehicle to accelerate. Since $F = m \times a$, we can write:

$$\text{k.e.} = m \times a \times s$$

Eventually, after being pushed this distance s, the vehicle will be moving at a velocity v.

v, a and s are all related by the equation of motion

$$v^2 = u^2 + 2as$$

In this case, $u = 0$ (the vehicle starts from rest) so

$$v^2 = 2as$$

or

$$as = \frac{v^2}{2}$$

so we can write our equation for kinetic energy as

$$\text{k.e.} = \tfrac{1}{2}mv^2$$

This, then, is how you calculate the kinetic energy of anything which is moving.

For example, if a cricket ball has a mass of 0.15 kg and a velocity of 20 m/s, its kinetic energy is given by

$$\begin{aligned}
\text{k.e.} &= \tfrac{1}{2}mv^2 \\
&= \tfrac{1}{2} \times 0.15\,\text{kg} \times 20\,\text{m/s} \times 20\,\text{m/s} \\
&= 30\,\text{J}
\end{aligned}$$

The kinetic energy of the cricket ball is 30 J.

7.7 Changing gravitational potential energy to kinetic energy

We have just found an equation that we can use to calculate kinetic energy. Is it right? Have we made a mistake somewhere? Should we not *test* this equation with an experiment? One of the ways in which scientists work is to think of what ought to happen, or to think up some equation, and then devise a way of testing their idea with an experiment. In this section you can do an experimental test to see if $\tfrac{1}{2}mv^2$ could be the right way to calculate kinetic energy.

Look at Fig. 7.9. When the mass is released it drops to the floor, pulling the trolley with it. The mass loses gravitational potential energy; this energy becomes kinetic energy of the trolley.

In this experiment, you must carry out the following.

1 Calculate the amount of gravitational potential energy lost by the falling mass. How will you do this?
2 Calculate the kinetic energy gained by the trolley, using the equation k.e. $= \tfrac{1}{2}mv^2$.
3 See if the two answers are the same. If they are, this would suggest that $\tfrac{1}{2}mv^2$ is indeed the right equation for calculating kinetic energy.

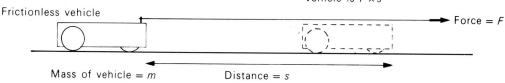

When the vehicle reaches here, the energy that has been transferred to it is $F \times s$, so the kinetic energy gained by the vehicle is $F \times s$

Frictionless vehicle

Force = F

Mass of vehicle = m

Distance = s

Fig. 7.8

You need to know the fastest velocity of the trolley (Why the fastest?) in order to work out its kinetic energy. You could find this from a ticker-tape record, as is being made in Fig. 7.9, or, if you prefer, you could devise an experiment in which the trolley cuts a light beam falling on a light-sensitive sensor, with a suitable electronic clock to measure the time for which the light beam was cut. (This technique will be used in the next section.)

The other measurements you need to make are shown in the table below. Carry out an experiment as suggested by Fig. 7.9 to make these measurements. Write your results on your copy of the table.

Table of results	
Mass of falling mass	kg
Height through which mass falls	m
Mass of trolley	kg
Highest velocity of trolley	m/s
Gravitational potential energy lost by falling mass (mgh)	J
Kinetic energy gained by trolley (assuming $\frac{1}{2}mv^2$ is right equation)	J

Do you think, from your results, that $\frac{1}{2}mv^2$ is the right equation for kinetic energy?

Do you think your experiment is a good one? Are you sure, for example, that *all* the gravitational potential energy became the kinetic energy of the trolley? Did anything else gain any kinetic energy? Can you think of any *improvements* to your experiment?

You may have noticed that we are assuming that *all* the gravitational potential energy must become *some* sort of energy; the energy does not disappear, nor can it just suddenly appear. This is what the **law of conservation of energy** is about.

7.8 Changing elastic potential energy to kinetic energy

Here is another energy change for you to look at—from elastic potential energy stored in the rubber band to kinetic energy of the air-track vehicle.

In Fig. 7.10 the tension in the stretched rubber band is 5 N, while Fig. 7.11 shows that the rubber has been stretched by 4 cm (0.04 m). What, then, was the elastic potential energy stored in the rubber?

Fig. 7.10   What is the tension in the elastic?

Fig. 7.9
What energy
changes take
place here?

Fig. 7.11   By how much has the elastic been extended?

In the next photograph the rubber band has just catapulted a vehicle along the air track. The vehicle has just passed a light beam which falls on a light-activated switch. This switch is connected to the VELA in front of the air track. The VELA is showing a time of 0.153 s, which is the time taken for the piece of black card on the vehicle to pass through the light beam.

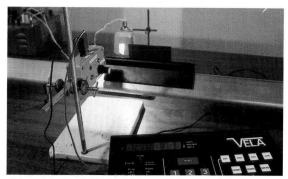

Fig. 7.12   How long did it take for the vehicle to pass the light beam?

As you can see from Fig. 7.13, the piece of black card was 10 cm (0.1 m) long. What, then, was the velocity of the air-track vehicle (in m/s)?

The mass of the vehicle was 365 g (0.365 kg) (see Fig. 7.14). What, then, was the kinetic energy of the vehicle, in joules? (Use k.e. = $\frac{1}{2}mv^2$.)

Is this answer the same as the elastic potential energy stored in the rubber band?

Fig. 7.13
How long is the piece of card?

Does it appear from these results that $\frac{1}{2}mv^2$ is the right way to calculate kinetic energy?

Do you think this is an accurate experiment? Can you think of any way of improving it?

7.9   Two numerical examples involving energy changes

Example 1

A tennis ball is dropped from a height of 1.8 m. It rebounds to a height of 1.25 m.
a) Describe the energy changes which take place.
b) With what velocity does the ball hit the ground?
c) With what velocity does the ball leave the ground?
Assume $g = 10$ N/kg.

Part a)
The energy changes are: gravitational potential energy of ball to kinetic energy of ball just before it hits the ground, to elastic potential energy in squashed ball while on the ground, to kinetic energy of ball as it leaves the ground, to gravitational potential energy of ball when it has risen to 1.25 m. Some of the ball's kinetic energy becomes internal energy of the ball and ground as it hits the ground.

Part b)
To solve this part you assume that all the gravitational potential energy of the ball becomes kinetic energy of the ball.

$$\text{g.p.e. of ball} = \text{k.e. of ball}$$
$$mgh = \frac{1}{2}mv^2$$
$$gh = \frac{1}{2}v^2$$

(Notice that the mass m cancels from both sides of the equation, so we do not need to know the mass of the ball.)

$$10 \text{ N/kg} \times 1.8 \text{ m} = \frac{1}{2}v^2$$
$$v^2 = 2 \times 10 \text{ N/kg} \times 1.8 \text{ m}$$
$$= 36 \text{ (m/s)}^2$$
$$v = 6 \text{ m/s}$$

The ball hits the ground at 6 m/s.

Fig. 7.14   What is the mass of the vehicle? (The round masses are each 100 g)

Part c)

After the ball leaves the ground, its kinetic energy becomes gravitational potential energy again. The question gives us information so that we can calculate this potential energy.

$$\text{k.e. of ball} = \text{g.p.e. of ball}$$
$$\tfrac{1}{2}mv^2 = mgh$$
$$\tfrac{1}{2}v^2 = 10\,\text{N/kg} \times 1.25\,\text{m}$$
$$v^2 = 2 \times 10\,\text{N/kg} \times 1.25\,\text{m}$$
$$= 25\,(\text{m/s})^2$$
$$v = 5\,\text{m/s}$$

The ball leaves the ground at 5 m/s.

Example 2

A car is travelling at 15 m/s (30 mph). When 25 m away from a road junction the driver applies the brakes. The mass of the car is 1 tonne. What average force must the brakes exert on the car in order that the car just stops at the road junction?

This problem can be solved using the equation

$$\text{force} \times \text{distance} = \text{change in energy}$$

In this case, *all* the kinetic energy is changed to another form of energy. (What sort of energy?) So

$$\text{force} \times \text{distance} = \text{change in kinetic energy}$$

$$
\begin{aligned}
\text{velocity} \quad & v = 15\,\text{m/s} \\
\text{distance} \quad & s = 25\,\text{m} \\
\text{mass} \quad & m = 1000\,\text{kg} \\
\text{force} \quad & F = ?
\end{aligned}
$$

Using
$$Fs = \tfrac{1}{2}mv^2$$
$$F \times 25\,\text{m} = \tfrac{1}{2} \times 1000\,\text{kg} \times 15\,\text{m/s} \times 15\,\text{m/s}$$
$$F = \frac{1000\,\text{kg} \times 15\,\text{m/s} \times 15\,\text{m/s}}{2 \times 25\,\text{m}}$$
$$= 4500\,\text{N}$$

The brakes apply an average force of 4500 N.

This question *could* have been solved by calculating the deceleration (How?) and then using $F = ma$ to calculate the braking force. This would have taken longer, with more chances for making mistakes! There is often more than one way to solve physics problems. Experience will help you to decide which is the best or quickest way.

7.10 Energy and friction

Fig. 7.15   What happened to all their energy?

The children shown in Fig. 7.15 soon felt exhausted as they transferred energy to the Land Rover. How would you find out how much energy they transferred to the Land Rover after pushing it 100 m?

What happened to all this energy? The Land Rover is not being lifted into the air so it did not gain any gravitational potential energy; it gained a little kinetic energy when it started moving, but it did not go on getting faster and anyway it stopped moving as soon as the pushing stopped. So where did the energy go?

You learned in Chapter 3 that whenever something moves there are likely to be some frictional forces; the children were pushing against frictional forces and the rolling resistance of the tyres (look back at Fig. 3.6). Whenever two surfaces rub against each other a little **internal energy** is produced, which will usually increase the temperature of the surfaces. This is what happened to all the energy the children transferred in Fig. 7.15.

The motor bike brake in Fig. 3.7 becomes hot for the same reason—the brake pads are rubbing against the brake disc. Brakes change the kinetic energy of a motor bike into internal energy in the brakes, thus increasing the temperature of the brakes. Engineers who design motor bikes and other vehicles must make sure there is a good flow of air past the brakes to cool them down, or else they could become too hot.

7.11 Elastic and inelastic collisions

Fig. 7.16   An inelastic collision!

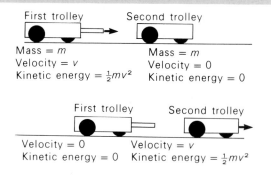

No kinetic energy is lost

Fig. 7.17   An elastic collision

Do any energy changes occur during a collision? In the collision which has just happened in Fig. 7.16 the answer must be 'yes', an energy transfer is needed to break the egg shell. Is there an energy transfer in *all* collisions.

You may already have examined simple collisions such as those outlined in Section 5.10. The simplest of those collisions involved a moving trolley hitting an identical trolley which was at rest. In this case the first trolley would stop and the second one would continue with the same velocity that the first one originally had. Fig. 7.17 illustrates this and calculates the kinetic energies of the trolleys. You can see that all the kinetic energy of the first trolley was transferred to the second trolley; no energy was transformed to a different kind. This kind of collision, in which no kinetic energy is lost, is called an **elastic collision**. This kind of collision is rare among large objects in everyday life, but there are millions of elastic collisions taking place at this moment between the air molecules around you.

A collision in which some kinetic energy *is* transformed to other kinds of energy is called an **inelastic collision**. The most extreme form of inelastic collision occurs when two things collide head on and stop. After such a collision there is no kinetic energy. *All* the kinetic energy is transformed into some other kind.

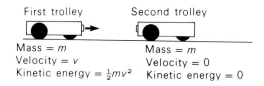

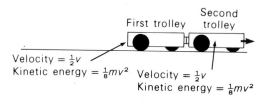

Total kinetic energy = $\frac{1}{4}mv^2$

Half the kinetic energy has been 'lost'

Fig. 7.18   An inelastic collision

SUMMARY

Now that you have finished studying this chapter on measuring energy, there are a number of things you should know and be able to do.

1 You should:
   a) know that energy transferred (or 'work done') is equal to $F \times s$;
   b) know that
      i) gravitational potential energy = mgh
      ii) elastic potential energy = $\frac{1}{2}Tx$
      iii) kinetic energy = $\frac{1}{2}mv^2$;
   c) know how to define 1 joule of energy;
   d) know that the area under a force/extension graph for a spring is equal to the energy stored in the spring;
   e) be able to carry out calculations involving energy transfers;
   f) know that frictional forces result in internal energy being produced.

2 You should know what each of the following is, or what each does:

elastic collision   inelastic collision
law of conservation   Hooke's law
    of energy

3 These are some of the other words that have been used in this chapter. You should know what each word means:

sufficient   virtually
technique

FURTHER QUESTIONS

1 Describe all the energy changes which take place in each of the following:
a) a swinging pendulum;
b) a tennis ball being hit with a tennis racket;
c) a hydroelectric power station producing electricity which lights a lamp.

2 Calculate the energy transferred (or work done) in each of the following cases:
a) a ship's anchor, of weight 8000 N, is lifted 25 m;
b) a 70 kg man climbs 4 m up a ladder;
c) a cable car of weight 4000 N is pulled up a mountain until it is 1000 m higher than when it started.

3 Calculate the kinetic energy of:
a) a trolley of mass 0.5 kg moving at 1 m/s;
b) a 1000 kg car moving at 20 m/s;
c) a 100 gram ball moving at 5 m/s.

4 A cricket ball of mass 160 g is hit vertically upwards with a velocity of 15 m/s.
a) How much kinetic energy does it have just after it is hit?
b) How much kinetic energy does it have at the highest point of its flight?
c) How much gravitational potential energy has it gained when it reaches the highest point of its flight?
d) How high does the cricket ball go?
e) How much kinetic energy does it have as it hits the ground?
f) What assumption have you had to make in answering parts c) to e)?
g) What happens to the ball's kinetic energy when it hits the ground?

5 A bicycle and rider start from rest and free-wheel down a hill. The bottom of the hill is 30 m below the top of the hill.
a) What is the maximum possible speed of the bicycle at the bottom of the hill?
b) In fact, the speed of the bicycle will be less than that which you have calculated above. Explain why.

6 A railway wagon of mass 5 tonnes is shunted at a speed of 2 m/s into four identical wagons. The wagons couple together.
a) What is the speed of the five wagons after they couple together?
b) What is the kinetic energy of the single wagon before the collision?
c) What is the total kinetic energy of the wagons after the collision?
d) How much kinetic energy has been 'lost'? What has happened to this energy?

7 A car has a mass of 1000 kg and is travelling at a speed of 20 m/s when the brakes are applied. It is brought to rest in a distance of 20 m.
a) How much kinetic energy has the car before the brakes are applied?
b) What was the average force exerted by the brakes?
c) What happens to the kinetic energy of the car?
d) What would be the stopping distance if the car was travelling at 40 m/s, assuming that there is the same braking force?

8 A building site has a ramp which is 6 m long along which workmen can push wheelbarrows. One end of the ramp is 1 m higher than the other. A workman has to push with a force of 250 N to push his wheelbarrow up the slope. The wheelbarrow has a mass of 100 kg.
a) How much work does the workman do in pushing his barrow up the slope?
b) How much gravitational potential energy does the wheelbarrow gain as it is pushed up the slope?
c) How much of the energy supplied by the workman has not been transferred to gravitational potential energy of the barrow? What has happened to this energy?
d) What is the value of the frictional force against which the workman must push when moving his barrow up the ramp?

8

ENGINES AND MACHINES

Fig. 8.1
Not a common sight

8.1   Introduction

Once, man just had the help of his horse to plough his fields. Energy was needed to turn over the soil, so the horse converted some of its store of chemical energy, which had come from its food, to supply what was needed.

Today, ploughing by horse in Britain is extremely rare; farmers use tractors. Why? The energy changes are very similar in both cases. A field needs much the same amount of energy to turn over its soil as it always did (the tractor-drawn plough does actually plough deeper than the horse-drawn plough), and the tractor converts chemical energy just as a horse does (although the chemical energy comes from a fossil fuel in this case). What is the advantage of a tractor?

Fig. 8.2
Why is this much more common than the scene in the last photograph?

One answer is to do with the time taken to do the job. A tractor might be able to plough a field in a few hours while a horse would take a week. Both will convert much the same amount of energy, but the tractor does so in much less time. The tractor is more *powerful*.

The industrial revolution started when people learned to make machines and engines to help them with their work. Engines, like the tractor engine, could work at a far greater rate than any person or animal. A so-called 'developed' country is one with many machines to do the work; a 'developing' country relies much more on people and animals. An ox still pulls a plough in many countries of the world.

In this chapter you will learn how some engines work, and how some machines enable people to exert far greater forces than they could without the machines.

8.2 You should know . . .

In order to understand this chapter, you should have read and understood Chapters 1, 2, 3, 4 and 7 in this book. In particular, you should understand how energy is measured (Chapter 7).

8.3 Measuring power

Even today we often compare the power of an engine with the power of a horse—the power of an engine is often measured in 'horsepower'. The tractor in Fig. 8.2 has a 50 horsepower engine. This means that, in a given time, it would take 50 horses to convert the same amount of energy as one tractor working at full power. Alternatively, we say that a tractor converts energy at the same *rate* as 50 horses.

Power is defined as the rate of transfer of energy, in other words, the amount of energy transferred every unit of time. It can be calculated from

$$\text{power} = \frac{\text{energy transferred}}{\text{time taken}}$$

Scientists do not usually use horsepower as their unit of energy. Since we measure energy in joules, and time in seconds, the units of power can be *joules per second*; these are usually called **watts** (after a British scientist called James Watt).

1 watt = 1 joule of energy transferred every second.

You will frequently find power measured in **kilowatts** (1000 watts), **megawatts** (1 000 000, or 10^6, watts) or even **gigawatts** (10^9 watts).

Example

What must be the power of the rescue helicopter's winch (Fig. 8.3) if it can lift an 80 kg person 30 metres in 15 seconds? Assume $g = 10$ N/kg.

Fig. 8.3   What is the power of the winch?

Answer

person's mass $m = 80$ kg
height gained $h = 30$ m
time taken $t = 15$ s
$g = 10$ N/kg

g.p.e. gained $= mgh$
$= 80\,\text{kg} \times 10\,\text{N/kg} \times 30\,\text{m}$
$= 24\,000\,\text{J}$

power = energy gained per second
$= \dfrac{24\,000\,\text{J}}{15\,\text{s}}$
$= 1600\,\text{W}$ (or 1.6 kW)

('W' is the abbreviation we use for watts.)

In practice, the power of the winch will need to be greater than we have just calculated; suggest some reasons why this is so.

8.4 Measuring various powers

How powerful are your legs and arms? How powerful is an electric motor you might find in your lab? Here are the outlines of some experiments you can do to find out. You will have to fill in some missing details and finish designing the experiments suggested here.

1 What power do your leg muscles develop as you run upstairs?

You need to know:

a) how much gravitational potential energy you gain when you run upstairs;

b) the time you take to run upstairs.

You must then use the equation

$$\text{power} = \frac{\text{energy gained}}{\text{time taken}}$$

What two measurements must you make in order to calculate the gravitational potential energy you gain when you go up the stairs?

How could you make these measurements?

In Fig. 8.4, a stopclock is being used to measure the time it takes for Toby to run upstairs. Do you think this is a sensible choice of timer?

After you have answered these questions, design and carry out an experiment along the lines that have been suggested so that you can calculate the power of your leg muscles.

Fig. 8.4 How powerful are Toby's legs?

2 What is the power of your legs when they are working a 'bicycle ergometer'?

A bicycle ergometer is the machine shown in Fig. 8.5. The piece of string rubs against the large wheel. The amount of friction created by this string can be altered by changing the mass hanging on the end of the string.

Fig. 8.5 Another way to measure the power of Toby's legs

What happens to all the energy that Toby is converting?

Look at Fig. 8.5. The mass hanging on the string weighs 50 N. The forcemeter at the other end of the string reads 10 N, so the forcemeter supports 10 N of the 50 N weight. The frictional force of the string against the wheel supports the other 40 N.

The circumference of the wheel of the bicycle ergometer is 1 m. How do you think this was measured?

Every time the wheel goes round once, the frictional force of 40 N acts for a distance of 1 m (round the circumference). How much work is done (energy is transferred) when the wheel goes round once?

How much work is done when the wheel goes round 50 times?

What other measurement must be made before the power of Toby's legs can be calculated?

If possible, carry out an experiment along the lines suggested above. Is the power of your legs when pedalling a bicycle ergometer similar to their power when running upstairs?

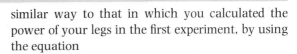

3 Measuring the power of your arm using an arm ergometer

The arm ergometer that Michael is using in Fig. 8.6 works in a similar way to the bicycle ergometer. Notice that the mass hanging on the end of the string is only 1 kg in this case. Why do you think it is less than the mass used in the bicycle ergometer?

Design an experiment using this apparatus to measure the power of your arms. Is your left arm as powerful as your right arm?

Fig. 8.6   Michael using an arm ergometer

4 Measuring the power of your arms by climbing a rope

If possible, try climbing a rope using your arms only (Fig. 8.7). Calculate the power of your arms in a similar way to that in which you calculated the power of your legs in the first experiment, by using the equation

$$\text{power} = \frac{\text{gain in g.p.e.}}{\text{time taken}}$$

Do your arms seem to be as powerful as when turning an arm ergometer?

The chemical energy converted by your body in the last four experiments is actually much larger than you have calculated. Only about one-fifth of the chemical energy converted in your muscles is actually used to increase your potential energy or drive the ergometers. What do you think happens to the other four-fifths?

This means that your body must actually convert the chemical energy in your muscles at about five times the rate you have calculated. We say your muscles are about $\frac{1}{5}$, or 20%, efficient.

5 Power of an electric motor

What energy changes are taking place in the apparatus shown in Fig. 8.8? Is this similar to any of the previous four experiments? What measurements must you make to calculate the power of the motor?

Carry out an experiment along the lines suggested by the photograph to find the power of an electric motor.

Fig. 8.7   Are Michael's arms just as powerful when measured this way?

Fig. 8.8   Measuring the power of an electric motor

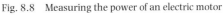

8.5 Energy and the steam engine

Fig. 8.9 Could all this energy be used?

There should be little doubt about the energy changes taking place in this photograph. What sort of energy is being produced? Where does this energy come from?

Your answer to the last question may well be: 'From the wood that is burning.' But how did the wood get its store of energy in the first place. Where did the energy *originally* come from?

How can you make use of the energy being formed by the fire? One of the earliest known ideas is 'Hero's steam engine' illustrated in Fig. 8.10. The hollow ball had water in it, which was heated by the fire underneath. The water boiled to make steam, which came out of the two jets. The ball was on an axle, so it was driven round by the jets of steam.

What energy changes were taking place?

Explain, *using momentum*, how the steam is able to turn the ball.

What modern engine works by using a similar physical principle?

Who was Hero? When did he live?

To Hero, this steam engine was just a toy. Nobody at the time thought of using the idea to do something *useful*. There was no need. Slaves were cheap and plentiful, and could do all the work that was needed.

It was not until the industrial revolution that serious efforts were made to use the energy from burning fuels to drive machines. The earliest examples were the engines used to pump water from the metal mines in Cornwall and the collieries in Durham and Northumberland. The pump invented by Newcomen was commonly used; the steam was fed into a cylinder where it condensed to water (see Fig. 8.11). This made the pressure in the cylinder very low, and atmospheric pressure pushed in the piston connected to the pump, which pumped water from the mines.

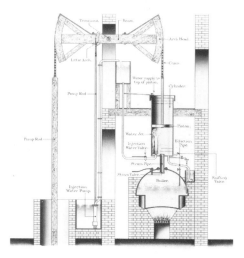

NEWCOMEN'S PUMPING ENGINE. 1712.

Fig. 8.11 Newcomen's steam engine

Fig. 8.10 Hero's steam engine

Newcomen's engine was very slow and cumbersome. It could not be made any more powerful because the materials available at that time were not strong enough. It was a Cornishman called Trevithick who developed the idea of a high-pressure steam engine in which the pressure of the *steam* pushed the piston. This enabled a greater power to be produced, and enabled Trevithick to construct the first steam locomotive to haul wagons in 1804. Although the materials available to Trevithick were much better than in Newcomen's time, Trevithick still had problems. His locomotive broke the rails on which it ran.

This is an example of what often happens in engineering. Successfully solving one problem results in many new problems to solve.

The steam locomotive was soon developed into a powerful machine which for over 100 years was the only means of moving heavy loads easily. Its place has now been taken by electric and diesel locomotives, which are cleaner and more efficient.

Fig. 8.12   Steam at work on the Festiniog Railway

Steam power is still widely used, not usually to drive a piston, as in a steam locomotive, but to drive a turbine. The electricity supply to your house and school was quite likely generated by a steam-driven turbine, the steam having been made by burning coal or oil, or from the heat of a nuclear reaction. (There is much more about electricity generation in Chapter 16.)

8.6   The petrol engine

Petrol and diesel engines are called **internal combustion engines** because the fuel burning takes place inside the engine, rather than outside it as in a steam engine. How does a petrol engine work?

Fig. 8.13 shows the inside of a petrol engine. A petrol engine usually has four, six or eight cylinders. Inside each cylinder is a piston which moves up and down turning a crank at the bottom of the cylinder. This crank drives the wheels of the car, via a gearbox. What makes the cylinder move up and down?

Look at Fig. 8.13 which shows the sequence of events inside a petrol engine. Study these diagrams and then answer these questions.

1 In diagram c), what energy changes are taking place?
2 Why do you think the gases in the cylinder expand [diagram c)] and push the piston down?
3 In diagram d), the piston is moving up and pushing the exhaust gases out of the exhaust valve. What is pushing the cylinder up? (Remember there is usually more than one cylinder and piston in a real engine.)
4 The cylinder is usually surrounded by a jacket of moving water? What do you think this water is for?

The engine illustrated is usually called a **four-stroke petrol engine**.

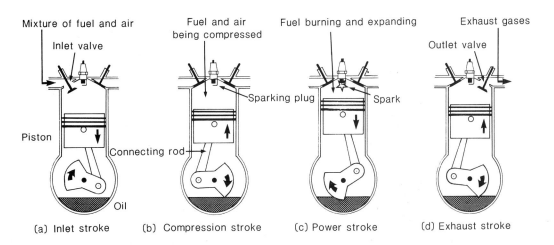

Fig. 8.13   The four stroke petrol engine

8.7 The diesel engine

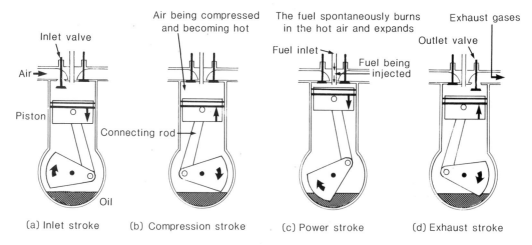

Fig. 8.14   The diesel engine

A diesel engine is constructed in a similar way to a petrol engine, but works in a different way. Fig. 8.14 shows details of the sequence of events in a diesel engine. You can see that during the induction stroke [diagram a)] only air enters the cylinder. How is this different from what happened in a petrol engine?

The air is compressed much more than the fuel/air mixture in a petrol engine—to about $\frac{1}{20}$ of the original volume instead of about $\frac{1}{10}$. As a result the air becomes very hot indeed. At the top of this stroke a squirt of diesel fuel enters the cylinder and this ignites in the hot air. The air becomes much hotter, the pressure increases and the piston is pushed down as in the petrol engine.

Diesel engines are more efficient than petrol engines, so they are more economical on fuel, but they need to be stronger so they are heavier and more expensive to make. Large, powerful engines for driving lorries, locomotives or ships are often diesel engines.

8.8   Levers

A crowbar, or lever, was probably one of the earliest 'machines' used by mankind. You may not think of the crowbar as a machine; to a physicist a machine transmits a force from one place to another; also a machine often makes a small force into a big one. A crowbar does both of these things, so we can call it a machine. What other machines have you met in this chapter?

Fig. 8.15   A crowbar in use!

Fig. 8.16   Why does Nicholas balance Judith?

How can a lever make a small force into a big one? How can Nicholas balance his heavier sister on the see-saw (which is just a large lever)? Notice that Nicholas is sitting much further from the pivot (or fulcrum) of the see-saw than Judith. Notice that the heavy rock is much nearer the pivot of the crowbar than the hand pulling the other end of the crowbar.

You have probably carried out experiments with levers before, perhaps with levers like the one Vicki is using in Fig. 8.17, trying to find a balancing rule. If not, now is the time to do so.

Fig. 8.17   Is Vicki about to balance this lever?

Fig. 8.18 shows a lever that is balanced. The 0.1 N force trying to turn the lever anticlockwise is balanced by the 0.3 N force trying to turn the lever clockwise. But the 0.3 N force is acting 0.2 m from the pivot, whereas the 0.1 N force is 0.6 m from the pivot. The product of force × distance is the same for both—0.06 N m. This is the **turning moment** of each force. For a lever to balance, the sum of all the moments trying to turn it clockwise must be equal to the sum of all the moments trying to turn it anticlockwise.

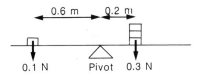

Fig. 8.18   Turning forces

Do you think Vicki, in Fig. 8.17, is about to balance her lever correctly?

To calculate a turning moment you must multiply the force by the distance from the pivot that the force acts. When you were calculating energy transfers earlier in this chapter you also multiplied force by distance. There is a very important difference between these two cases. When calculating energy transfers, the force and the distance were both in the *same direction*; when calculating moments the direction of the force is at *right angles* to direction to the pivot. The units of a turning moment are newton metres, *not* joules.

Suppose the rock being moved by the crowbar in Fig. 8.19 weighs 2000 N, and the crowbar is pivoted 0.2 m from the rock. How much force must be exerted to just move the rock, if the force on the crowbar is applied 1 m from the pivot?

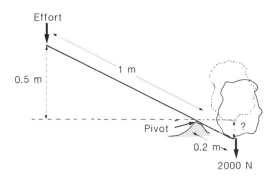

Fig. 8.19

The anticlockwise moment of the system is

$$\text{moment} = \text{force} \times \text{distance from pivot}$$
$$= 2000 \, \text{N} \times 0.2 \, \text{m}$$
$$= 400 \, \text{N m}$$

To just move the rock, the clockwise moment must also be 400 N m.

$$\text{moment} = \text{force} \times \text{distance from pivot}$$
$$400 \, \text{N m} = F \times 1 \, \text{m}$$
$$F = 400 \, \text{N}$$

A force of 400 N is required.

We say that the force of 2000 N is the **load** on this lever, and the force of 400 N is the **effort**. In this case, the load is five times greater than the effort, so the **mechanical advantage** of the system is 5.

$$\text{mechanical advantage} = \frac{\text{load}}{\text{effort}}$$

Moving a big force with a little one seems suspiciously like getting something for nothing! Inevitably, there is a price to pay. Suppose the crowbar is pulled down 0.5 m. How far up will the rock move? (Fig. 8.19 might help.) How much gravitational potential energy does the rock gain? How much energy is transferred when pulling the crowbar down 0.5 m? Was there any gain of energy when using the crowbar?

We have assumed in this section that there are no frictional forces to overcome. In practice, it will be necessary to pull with a force of slightly greater than 400 N because of the friction on the crowbar at the pivot.

8.9 Pulleys

Fig. 8.20   A pulley system at work

Fig. 8.22   Richard investigating a pulley system

The forcemeter that Richard is using showed that an effort of 3 N was needed to raise the pulley and its 5 N load. What is the mechanical advantage of this system?

Suppose the 5 N load was lifted 10 cm. The gain in gravitational potential energy of the load is given by

$$\text{gain in g.p.e.} = \text{weight} \times \text{height gained}$$
$$= 5 \,\text{N} \times 0.1 \,\text{m}$$
$$= 0.5 \,\text{J}$$

If the load moves 10 cm, we agreed above that Richard's hand ('the effort') must move 20 cm, so the work done by Richard is given by

$$\text{work done} = \text{force} \times \text{distance}$$
$$= 3 \,\text{N} \times 0.2 \,\text{m}$$
$$= 0.6 \,\text{J}$$

Richard transfers 0.6 J of energy to the pulley system, but the load only gains 0.5 J of energy. What happens to the other 0.1 J of energy? Answer: the pulley itself gains some gravitational potential energy, and a little internal energy is formed as a result of friction in the pulley's bearings.

We say that the machine is not 100% **efficient** at transferring the 'input' energy to useful 'output' energy. We *measure* efficiency using the following equation:

efficiency

$$= \frac{\text{useful energy obtained from machine}}{\text{energy supplied to machine}} \times 100\%$$

in the case of this pulley system,

$$\text{efficiency} = \frac{0.5}{0.6} \times 100\%$$

$$= 83\%$$

Since you get energy out of the pulley system (or any machine) at the same time as you supply

Like the crowbar in the last section, the pulley system on this crane is a 'force multiplier'; a large load can be lifted with a small effort. As before, the load does not move as far as the effort.

Look at the simple pulley system in Fig. 8.21. To raise the pulley and its load 10 cm, 20 cm of string must be pulled through the pulley, since the bit of string on the left of the pulley and the bit of string on the right of the pulley must both be 10 cm shorter. The effort that Richard applies must move *20 cm* to move the load *10 cm*, so the effort moves twice as fast as the load; we say that the **velocity ratio** of the system is 2.

$$\text{velocity ratio} = \frac{\text{distance effort moves}}{\text{distance load moves}}$$

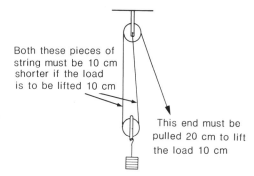

Both these pieces of string must be 10 cm shorter if the load is to be lifted 10 cm

This end must be pulled 20 cm to lift the load 10 cm

Fig. 8.21   A simple pulley system

energy, it is also true to say that

$$\text{efficiency} = \frac{\text{useful power obtained}}{\text{power supplied}}$$

You have already met the idea of efficiency in Section 8.4 where you were using a bicycle ergometer.

If the pulley system in Fig. 8.22 were 100% efficient, Richard would only need to supply an effort of 2.5 N. The mechanical advantage would then be 2, the same as the velocity ratio. No machine is 100% efficient in practice, although the crowbar in the last section very nearly is; look back at the last section and you will see that we did assume that *all* the energy supplied became gravitational potential energy of the rock.

It is certainly impossible for any machine to be more than 100% efficient. Why?

Here is a quick way to work out the efficiency of a machine if you happen to know the velocity ratio and the mechanical advantage:

$$\text{efficiency} = \frac{\text{useful energy out}}{\text{energy in}}$$

$$= \frac{\text{load} \times \text{distance load moves}}{\text{effort} \times \text{distance effort moves}}$$

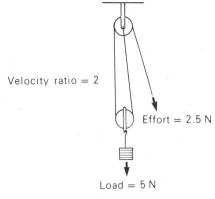

Velocity ratio = 2

Effort = 2.5 N

Load = 5 N

Mechanical advantage = 2

Efficiency = 100%

Fig. 8.23

$$= \text{mechanical advantage} \times \frac{1}{\text{velocity ratio}}$$

i.e.

$$\text{efficiency} = \frac{\text{mechanical advantage}}{\text{velocity ratio}}$$

You should now be able to investigate more complicated pulley systems, like the one on the crane in Fig. 8.20, to make sure that all these ideas apply to more complex systems.

SUMMARY

Now that you have finished studying this chapter on engines and machines, there are a number of things you should know and be able to do.

1 You should:
   a) know that power is energy transferred per unit of time;
   b) know the definition of 1 watt of power;
   c) know what a turning moment is, and how to calculate one;
   d) know that the units of turning moment are newton metres (N m);
   e) understand that, for a lever to balance, the clockwise turning moments must equal the anticlockwise turning moments;
   f) know what is meant by, and how to calculate
      i) mechanical advantage
      ii) velocity ratio;
   g) understand why machines are never 100% efficient;
   h) be able to calculate the efficiency of a machine, using either

$$\text{efficiency} = \frac{\text{energy supplied}}{\text{useful energy out}}$$

$$\text{or} \quad \text{efficiency} = \frac{\text{useful power out}}{\text{power in}}$$

$$\text{or} \quad \text{efficiency} = \frac{\text{mechanical advantage}}{\text{velocity ratio}}$$

   i) be able to explain how both petrol engines and diesel engines work.

2 You should know what each of the following is, or what it does:

arm ergometer	pivot
load	crowbar
fulcrum	machine
lever	effort
pulley	engine
locomotive	crank
internal combustion engine	
bicycle ergometer	

3 Here are some of the other words that have been used in this chapter. You should know what each word means:

virtually   straightforward   cumbersome   sequence
frictionless   colliery   construct   convert
                                            compress   complex
                                            balance   system
                                            product   economical

FURTHER QUESTIONS

1 When the photograph which forms Fig. 8.4 was taken, Toby, who has a mass of 60 kg, ran up the flight of stairs in 5 seconds. The difference in height between the bottom and top of the stairs is 3 m. When Kate, who has a mass of 45 kg, ran up the same stairs, she took 4 s.
   a) How much gravitational potential energy did Toby gain?
   b) How much gravitational potential energy did Kate gain?
   c) Who developed the greater power in their legs? What was that power?
   d) How much chemical energy did each person convert if the efficiency of their leg muscles is 20%?

2 An electric motor with an output power of 20 kW lifts a 500 kg crate 20 m to the top floor of a warehouse.
   a) How much gravitational potential energy does the crate gain?
   b) How long does it take for the motor to lift the crate to the top of the warehouse?

3 The Niagara Falls are 50 m high. It has been estimated that an average of 7×10^6 kg of water pours over the Falls every second.
   a) If all the energy of the falling water could be converted to electrical energy, what electric power would be available?
   b) Suggest some reasons why it would not be possible to convert *all* the energy of the falling water to electrical energy.

4 A cyclist with a mass of 50 kg rides a bicycle of mass 20 kg up a hill. The top of the hill is 30 m higher than the bottom of the hill.
   a) What is the least power the cyclist must develop in order to ride up the hill in one minute?
   b) In practice, the cyclist would have to develop more power than this minimum power that you have just calculated. Explain why.

5 A crew of eight people are rowing a racing boat. Each person makes 30 strokes of their oar in a minute, and the length of that pull is 1 m. The average force with which each person pulls is 200 N.
   a) How much energy does one person transfer to the boat during one stroke of that person's oar?
   b) What is the average amount of work that one person does in 1 second?
   c) What is the average amount of work that all eight people do in 1 second?
   d) What is the average power of the crew?
   e) If, on average, the efficiency of the crew's muscles is 25%, what is the total amount of energy they need in order to row a race which lasts 5 minutes?

6 Why do you think machines are generally not 100% efficient?

7 Fig. 8.24 shows a system of pulleys. A 'load' of 20 kg is placed on the lower pulley. An effort of 100 N is required to just lift this load.
   a) If the load moves up 1 m, how much gravitational potential energy does the load gain?
   b) How far down must the effort move in order to lift the load 1 m? (Each of the three strings between the pulley blocks must shorten by 1 m.)
   c) How much energy is transferred from the effort to the pulley system when the effort moves the distance you have just calculated?
   d) What is the efficiency of the pulley system?
   e) Give two reasons why the system is not 100% efficient.

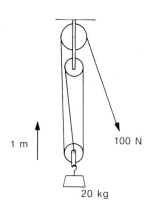

1 m   100 N

20 kg

Fig. 8.24

9

ENERGY AND MOLECULES

Fig. 9.1

Fig. 9.2

Figs 9.1 – 9.3
These are the three states of
water

Fig. 9.3

9.1 Introduction

Ice, water and steam are the three *states* of water. The same kind of molecules are present in each case but they are put together in ways which give quite different results. How are the molecules arranged in solids, liquids and gases?

Why do you need energy to boil water, that is, to change it from a liquid into a gas? Is this energy 'stored' in the steam in some way, or is it lost, or what? Can you get this energy back again, for example, when the steam condenses? You need energy to change ice into water, but how much energy?

You need energy to make anything warmer. What do the molecules of a substance do when the substance becomes warmer?

This chapter is about the effect of energy on the molecules in a substance, and helps you to answer these questions.

9.2 You should know . . .

In order to understand this chapter, you should:
1 know the difference between atoms and molecules;
2 have seen some of the evidence for the existence of atoms and molecules, particularly the evidence provided by crystals, Brownian motion and diffusion;
3 have some ideas of the way in which atoms and molecules are arranged in solids, liquids and gases;
4 know how to use a thermometer.

9.3 Particles in solids, liquids and gases

Fig. 9.4   Crystals of potassium nitrate

What happens to the particles in a substance when it melts, or when it boils? You know that the particles in a solid are arranged in a regular way, and this is the reason for the regular shape of the crystals in Fig. 9.4. Fig. 9.5 is a *model* showing how we think the particles of a solid (copper atoms in this case) are arranged. Each polystyrene sphere represents an atom. Why is it useful to have models when discussing atoms and crystals?

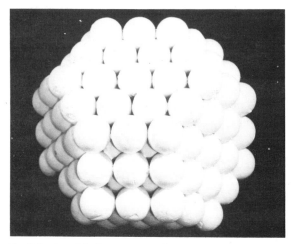

Fig. 9.5   A model of a copper crystal

This model shows that each atom has its own fixed place in the solid; it is tightly bonded to its neighbour. But this model does *not* show one important feature of solids. In solids, each atom vibrates, just a little. They vibrate in all directions, at random. The more energy that a solid has, the greater are these random vibrations, and hence the more kinetic energy each atom has. In a hot substance the particles have more random vibrational energy than in a cold substance.

As a solid becomes hotter, there comes a point where each atom has sufficient energy to tear itself away from its neighbours and move somewhere else—the solid has melted and become a liquid. How would you use the same polystyrene spheres as in Fig. 9.5 to make a model of a liquid? Do you have to keep your model in a container, like a real liquid?

Although your model probably does not show it, the particles in a real liquid are moving around, in random directions, all the time. As with a solid, the hotter the liquid, the faster the particles move. If the particles have enough energy, they can tear themselves away from each other; the substance becomes a gas, and the particles fly around in random directions at high speeds.

This has been a simple description of melting and boiling from a particle's point of view. The next few sections look at some of the points in this description in a little more detail.

9.4 Melting and freezing

Fig. 9.6   Why will this snow not all melt at once?

Have you ever noticed how long ice and snow remain on the ground in the winter, even after the weather becomes warmer? Even a 'sudden thaw' takes time; the snow and ice does not all melt at once. Why not? Conversely, why does a pond not all freeze as soon as the air temperature drops below 0 °C?

Fig. 9.7 shows an electric immersion heater in a filter funnel full of ice. The ice is gradually melting. The more energy that is supplied by the heater, the more ice melts. Energy is needed to pull the water molecules away from each other so that they are free to move around. The temperature does not change—both the ice and the water are at 0 °C. The greater the quantity of energy supplied, the greater the number of molecules which can be made free, and the more the ice melts.

Fig. 9.7 One way to supply energy to melt ice

Fig. 9.8 Investigating the cooling of stearic acid

If energy is needed to *melt* ice, or any other solid, what happens to this energy when water, or any other liquid, freezes again? This problem is quite easy to investigate using stearic acid, a white solid which melts at about 50 °C.

One-third fill a boiling tube with stearic acid and melt the stearic acid by heating the boiling tube gently in a Bunsen flame. When the stearic acid is molten, turn off the Bunsen burner and put the test tube with the molten stearic acid in a suitable holder as Kate has done in Fig. 9.8. Use a suitable thermometer to measure the temperature of the stearic acid at, say, half minute intervals until the stearic acid has all solidified. Record your results in a table and then draw a graph of the temperature of stearic acid against time.

Fig. 9.9 shows the graph which Kate obtained. At the start of the experiment the stearic acid had a high temperature and was cooling down, passing energy to the air surrounding the boiling tube. Then the graph levels off for several minutes. What was the stearic acid doing at this stage? At what temperature did this occur?

The stearic acid is still warmer than the surrounding air, so the air is still getting energy from the stearic acid, even though the stearic acid is not becoming any cooler. As the stearic acid freezes, it gives out the energy that was originally required to melt it. Solid stearic acid at 50 °C has less internal energy than the same quantity of liquid stearic acid at the same temperature. The energy the solidifying stearic acid gives out is called its **energy of fusion**. (To 'fuse' means to 'melt'.)

After some minutes the stearic acid continues to cool again. What has happened to it so that it can now cool down again?

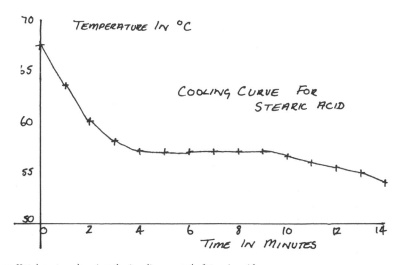

Fig. 9.9 A graph from Kate's notes, showing the 'cooling curve' of stearic acid

9.5 Specific energy of fusion

The apparatus in Fig. 9.7 can be used to measure *how much* energy is needed to melt a given mass of ice. The immersion heater is connected to the electricity supply via a joulemeter; this measures the total quantity of electrical energy supplied. The amount of ice which melts to form water can be measured by weighing the beaker before and after the experiment, so you will know how much energy (measured on the joulemeter) is needed to melt this mass of ice.

Physicists find it convenient to refer to the quantity of energy needed to melt *1 kilogram* of ice (at 0 °C). This is called the **specific energy of fusion** of ice.

Carry out the experiment outlined above and from your results calculate the specific energy of fusion of ice. There is an outline of a table for your results below to help you. If you do not have a joulemeter, you will have to measure the electrical energy using a voltmeter and ammeter. In this case, this experiment would be better done after you have read Chapter 14.

The apparatus shown in Fig. 9.7 will give a very inaccurate result for the specific energy of fusion of ice. Why? (Will any ice melt *without* the heater switched on?) Can you think of any ways of improving this experiment to give a more accurate result?

Table of results	
Reading of joulemeter at start	
Reading of joulemeter at finish	
Number of joules of energy used	
Mass of empty beaker	
Mass of beaker + melted ice	
Mass of ice which melted	

Calculation

_____ J of energy are needed to melt _____ kg of ice.

Hence _____ J of energy are needed to melt 1 kg of ice.

The specific energy of fusion of ice is _____ J/kg.

9.6 Evaporation and boiling

Fig. 9.10   How do these clouds form?

How do the clouds come to be in the sky? The air contains some water vapour; if the air becomes cooler this water vapour might condense to make tiny drops of water which collect together to form clouds. If the drops of water are small enough, currents of air keep them in the air, but if the water drops join together to form larger drops, they fall as rain.

The water vapour gets into the air in the first place because it *evaporates* from the surface of the sea, lakes or rivers. Some of the molecules on the surface of any liquid may have enough energy to escape from the liquid and form a vapour. Notice that evaporation takes place from the *surface* of a liquid.

If a liquid is hot enough, it boils. Boiling takes place throughout the whole volume of the liquid. Look at some water boiling; you can see bubbles of steam forming right at the bottom of the water. Boiling occurs when the pressure of the vapour leaving the liquid (the **vapour pressure**) is equal to the pressure of the air pushing down on the surface of the liquid.

Just as energy is needed to melt a solid without changing its temperature, energy is also needed to change a substance from liquid to gas without a change of temperature. The energy required to change 1 kilogram of a substance from liquid to gas (at a temperature equal to the boiling point) is called the **specific energy of vaporization**.

A refrigerator relies on energy of vaporization. Look at Fig. 9.11. A liquid (such as Freon) is allowed to boil (by reducing the pressure—see Section 9.8); the energy required comes from the inside of the refrigerator, thus cooling down the inside. The vapour is carried by a pipe to outside the body of the refrigerator, where it condenses again. When condensation occurs, the energy of vaporization is given out to the atmosphere. A refrigerator is an example of a heat pump.

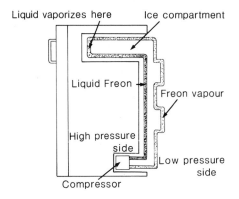

Liquid vaporizes here   Ice compartment
Liquid Freon
Freon vapour
High pressure side
Low pressure side
Compressor

Fig. 9.11   A refrigerator

The fastest particles will possibly have enough energy to tear themselves away from the surface of the liquid altogether—they will **evaporate**. Since the fastest particles leave the liquid, the average speed of the remaining particles is less. It is the slower ones that are left behind, so the temperature of the liquid goes down. The ethanol feels cold.

A cup of coffee cools down partly as a result of evaporation; the faster (hence hotter) molecules of water evaporate, leaving behind the slower (cooler) ones.

Bottles of milk can be kept cool without a refrigerator by covering them with a damp cloth. Explain, using the fact that water will evaporate from the cloth, how this keeps the milk cool.

Evaporation takes place faster if the molecules are blown away as they evaporate, they are then less likely to condense back into the liquid. You know that washing dries better on a windy day than on a still day, even if it is a cool day.

Why is a mountaineer who is wearing wet clothes in greater danger on a windy day than a companion whose clothes are dry?

9.7   Cooling by evaporation

Put a drop of ethanol on the back of your hand. Why does it feel cold? (You may have experienced the same effect when having ethanol wiped on your skin prior to an injection.)

You have learned that the temperature of a liquid (ethanol in this case) depends on how fast the particles are moving around; the faster they move, the warmer the liquid. However, not all the particles in a liquid are moving at the same speed; some are moving very slowly and others are moving very fast. Most are moving at a speed between these two extremes, as shown in Fig. 9.12. It would be more accurate to say that the temperature of the liquid depends on the *average* speed of the particles.

Fig. 9.13
Evaporation helps make this cup of coffee cool down

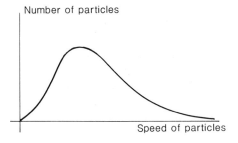

Number of particles

Speed of particles

Fig. 9.12   Particles of a liquid have a spread of speeds

Fig. 9.14   The wind helps evaporation

9.8 Changing melting and boiling points

Fig. 9.15 Spreading salt on icy roads

Why is salt spread on icy roads in the winter? Why does the sea rarely freeze even though rivers and streams at the same temperature may have a surface of ice?

To answer these questions, get some ice from a freezer so that its temperature is well below 0 °C. Crush it and put it in a boiling tube with a little water. Allow the temperature to become steady and then measure the temperature of the water—it should be 0 °C.

Fig. 9.16
Does the temperature change if salt is added to the water?

Add a spatula measure of salt, stir it until it is dissolved in the water and then measure the temperature again. Is it the same as before? Dissolve another spatula measure of salt in the water. What happens to the temperature?

Anything dissolved in the water will have the same effect as the salt. Motorists add an **antifreeze** (such as glycol) to the water in their car engines to prevent the water freezing during the winter. If the water in the engine freezes, why might it cause damage to the engine?

As a result of your observations, explain why salt is put on the roads in icy weather.

Another way of changing the melting point of ice is shown in Fig. 9.17. The copper wire appears to be cutting its way through the block of ice. In fact, the wire is melting the ice just under the wire because the wire is exerting a large pressure on the ice. If the pressure on a solid increases, the melting point decreases. You can see that the water has frozen again above the wire where there is no increased pressure.

Fig. 9.17
Cutting through a block of ice with a copper wire; the ice freezes again behind the copper

Ice skating is helped by this fact. The pressure on the ice under a skater's boot melts the ice a little so that there is a thin film of water on top of the ice. This makes the friction between the boot and the ice very small. There is quite a lot of friction on the surface of absolutely dry ice.

Both impurities and pressure change the boiling point of a liquid. Every cook knows that the boiling point of water with sugar dissolved in it is higher than that of pure water, and that the greater the concentration of sugar, the higher the boiling point.

Fig. 9.18   Ice skating

The effect of pressure on the boiling point of water can be shown using the apparatus illustrated in Fig. 9.19. The water in the round-bottomed flask is boiled for a minute so that the steam drives out all the air in the flask. The Bunsen burner is removed and at the same time the clip on the rubber tubing is fastened tightly. There is a thermometer inserted through the stopper to measure the temperature of the water in the flask. The flask is held under running cold water. This condenses the steam in the flask, so the pressure goes down. The water will start to boil again, even though the temperature is now below 100 °C. By constantly condensing the steam that is formed, it is possible to make the water in such a flask boil at a temperature below 60 °C.

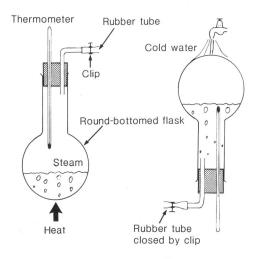

Fig. 9.19   Making water boil below 100 °C

It is said that tea made near the tops of high mountains does not taste very nice. The atmospheric pressure near the tops of such mountains is much less than the pressure at sea level, and water boils at much less than 100 °C. This is not hot enough to make a decent cup of tea.

The opposite effect occurs in a pressure cooker (Fig. 9.20). Here the lid is clamped tightly in position so that the steam cannot escape, so the pressure increases (up to a limit set by the safety valve). Water in this pressure cooker boils at about 110 °C, so the food in it will cook much more quickly than in a normal pan.

Fig. 9.20   Water boiling above 100 °C

9.9   Expansion

Both the next two photographs show something that is probably familiar to you—a gap in a road at one end of a motorway flyover, and a gap between two lengths of railway line. You probably know that these gaps are there to allow room for the bridge and railway line to expand when they become hot. Why do most things expand when the temperature increases?

Fig. 9.21
An expansion gap in a road bridge

Fig. 9.22
An expansion gap between railway lines

Remember that as a solid warms up the particles vibrate more. The result of this is that the average space between particles increases a little, so the solid itself becomes a little larger. Notice that the particles themselves do not become larger—it is the spaces between the particles that increase slightly.

Expanding bridges and railway lines are a nuisance. Bridges must be able to move on their supports. The gaps between railway lines cause extra wear and tear to the rolling stock. To overcome this last problem, many railway lines are now laid in continuous lengths of a kilometre or more. As they are laid they are stretched slightly (put under tension). If they become hotter this just reduces the tension that was put there when they were laid.

How big must the gap be between the railway lines in Fig. 9.22? It was a fairly cool day when the photograph was taken—about 10 °C. It is unlikely that the rail will become hotter than 50 °C, even with the Sun shining directly on it. Although you cannot see all of it, the length of each piece of rail in the photograph is about 20 m. By how much will 20 m of steel rail expand when heated from 10 °C to 50 °C, i.e. by 40 °C?

To answer this sort of question, engineers and scientists need to know the **linear expansivity** of a material. This is the amount by which *1 metre* of the material expands when its temperature increases by 1 °C. The linear expansivity of steel is 0.000 011/°C. In other words, 1 m of steel expands by 0.000 011 m if its temperature increases by 1 °C.

A 20 m length of steel will expand by 20 times as much, i.e.

$$20 \times 0.000\,011\,\text{m} = 0.000\,22\,\text{m}$$

If 20 m of steel increases its temperature by 40 °C instead of 1 °C, it will expand by a further 40 times as much, i.e.

$$40 \times 0.000\,22\,\text{m} = 0.008\,8\,\text{m} \quad (\text{or } 8.8\,\text{mm})$$

Does the gap in the rail in Fig. 9.22 look as if it is big enough?

Expansion is not always a nuisance. Fig. 9.23 shows one useful application of expansion. The steel tyre is heated until it has expanded sufficiently for the centre of the wheel to fit inside it. As the tyre cools, it contracts and grips the wheel centre firmly.

Each material has its own particular linear expansivity. Brass, for example, has a larger linear expansivity than steel. This fact is made use of in a **bimetallic strip** as illustrated in Fig. 9.24. The strip of brass is firmly riveted to the strip of steel. As the strip is heated, the brass expands more than the steel. What do you think will happen to the strip?

Fig. 9.23   Expanding a steel tyre by heating it to fit it to a wheel

Fig. 9.24   A bimetallic strip

Fig. 9.25 shows the bimetallic strip forming part of a circuit designed to switch on a central heating pump when the temperature drops below a certain value. How do you think the circuit works? What do you think screw 'S' is for?

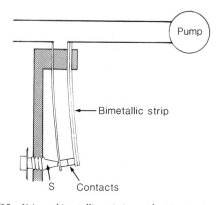

Fig. 9.25   Using a bimetallic strip in an electric circuit

Liquids also expand when heated. The mercury-in-glass thermometer that you have probably used during this topic makes use of this fact.

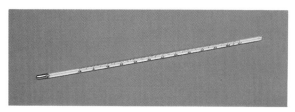

Fig. 9.26   A mercury thermometer relies on expansion of the mercury

9.10   Heating water and aluminium

Fig. 9.27   Whose 'billy' will boil first?

Fig. 9.28   Investigating the warming of water

Who is going to have a cup of tea first? Both primus stoves have the same power, that is, give the same heat energy per second to the billies of water, but you would expect the larger quantity of water to take longer to boil. The greater the mass of the water, the more heat energy is needed to increase the temperature by a given amount.

If there is *twice* as much water in the large billy as in the small one, will it need twice as much energy to increase its temperature to boiling point? Will it take twice as long to boil?

Fig. 9.28 shows Sheila performing a simple laboratory experiment to try to answer these questions. The can contained 1 kg of water which was heated by an electric immersion heater.

The temperature of the water was measured at 1 minute intervals for 20 minutes. Sheila then drew a graph of the temperature of the water against time.

She then repeated the experiment using 0.5 kg of water, drawing a graph of the results of the same axes as the first graph. Sheila's graphs are shown in Fig. 9.29.

One obvious feature of both graphs is that, although the temperature increases fairly steadily to start with, the lines eventually become horizontal at a temperature of about 30°C. Neither can of water ever boiled. Why not? What was happening to all the energy from the immersion heaters if it was not warming the water any more? How could Sheila improve this experiment so that the heaters continued to warm the water?

Now try the experiment, with your suggested improvements, for yourself.

As far as you can tell from the first part of the graphs in Fig. 9.29, does the 1 kg of water take twice as long as the 0.5 kg for its temperature to increase from 20°C to 25°C? Does this mean it needs twice as much energy? Do your graphs show the same?

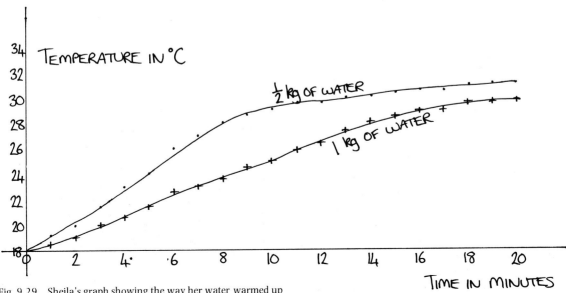

Fig. 9.29   Sheila's graph showing the way her water warmed up

Suppose Sheila had used a more powerful heater in this experiment. In what ways would you expect the graphs to be different from those in Fig. 9.29?

The temperature of the aluminium can which contained the water also increased; this needs energy. Does 1 kg of aluminium need the same quantity of energy as 1 kg of water to increase in temperature by 1 °C? To answer this question, a piece of aluminium into which an immersion heater will fit can be used, as in Fig. 9.30. The temperature can be measured at regular intervals as in the last experiment.

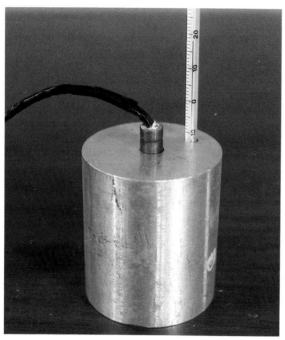

Fig. 9.30   A heater fitted in an aluminium block

Fig. 9.31 shows a graph prepared in the same way as before. Compared with 1 kg of water, does 1 kg of aluminium take more time or less time to increase its temperature by 1 °C? Does it need more energy or less energy?

9.11   Specific heating energy

From reading the last section it must be obvious to you that the energy needed to warm something up depends on what substance is being warmed, what mass of substance there is and what the change in temperature is.

Scientists find it convenient to refer to the energy needed to increase the temperature of *1 kilogram* of a substance by *1 °C*; this is known as the **specific heating energy** of the substance. (An alternative name is the **specific heat capacity** of the substance.) For example, the specific heating energy of water is 4200 J/kg °C. This means that 4200 J of energy are needed to increase the temperature of 1 kg of water by 1 °C. Similarly 8400 J are needed to increase the temperature of 1 kg of water by 2 °C, or to increase the temperature of 2 kg of water by 1 °C, and so on.

How much energy is needed to increase the temperature of:

a) 2 kg water by 2 °C;
b) 3 kg water by 1 °C;
c) 2 kg water by 4 °C;
d) 5 kg water by 3 °C;
e) 9.5 kg water by 0.1 °C?

You should be able to see from these examples that the energy required can be calculated from

energy = mass × specific heating energy
$$\times \text{ temperature change}$$

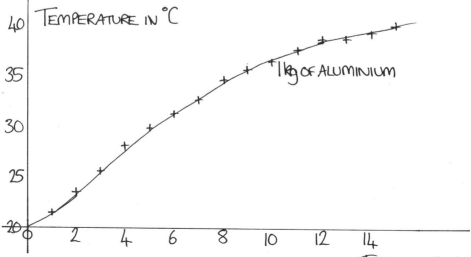

Fig. 9.31   How does this graph for the warming of aluminium compare to the last graph?

Look back at Fig. 9.31. Do you think the specific heating energy of aluminium is *more* or *less* than that of water? Roughly *how much* more or less?

How do we *know* that the specific heating energy of water is 4200 J/kg °C? To measure the specific heating energy of something you must measure how much energy is needed to increase the temperature of 1 kg of the substance by 1 °C. How can you obtain a known quantity of energy? The easiest and most accurate way is to use an electric immersion heater and measure the electrical energy which is converted to internal energy of the substance, thus increasing its temperature. You will find this described in Chapter 14, which is about electrical energy.

Equipment like the arm ergometer (see Section 8.4) could, in principle, be used to measure the specific heating energy of the ergometer wheel. You have to assume that all the energy you supply becomes internal energy in the wheel (as a result of friction between the string and the wheel). If you measure the mass and the temperature increase of the wheel, you can calculate the specific heating energy (using the equation above). Such an experiment is not likely to be very accurate. Why not? What possible sources of error are there?

9.12 A numerical example involving specific heating energy

A kettle has a 3 kW heating element and holds 1.5 kg of water. The specific heating energy (SHE) of water is 4200 J/kg °C.
a) The kettle is filled with water at 20 °C. How much energy is needed to bring the water to the boil?
b) How long will it take to boil the water?
c) What did you have to assume to answer part b)?

Part a)

$$\text{mass of water} = 1.5 \, \text{kg}$$
$$\text{temperature change} = (100-20)\,°C = 80\,°C$$
$$\text{SHE of water} = 4200 \, \text{J/kg}\,°C$$

$$\text{energy} = \text{mass} \times \text{SHE} \times \text{temperature change}$$
$$= 1.5 \, \text{kg} \times 4200 \, \text{J/kg}\,°C \times 80\,°C$$
$$= 504\,000 \, \text{J} \quad (\text{or } 504 \, \text{kJ})$$

504 kJ is needed to bring the water to the boil.

Part b)
The heating element supplies 3000 J of energy every second.

Time taken for water to boil

$$= \frac{\text{total energy required}}{\text{energy supplied/second}}$$

$$= \frac{504\,000 \, \text{J}}{3000 \, \text{J/s}}$$

$$= 168 \, \text{s}$$

It takes 168 seconds to boil the water.

Part c)
We had to assume that all the energy supplied by the heating element went to heating the water. In practice, some energy will be lost to the surrounding air and it will take longer than 168 seconds to boil the water.

SUMMARY

Now that you have finished studying this chapter on energy and molecules, there are a number of things you should know and be able to do.
1 You should:
  a) begin to understand why scientists sometimes use models, and know that all models have some limitations;
  b) know the ways in which particles are arranged in solids, liquids and gases;
  c) know why energy (the energy of fusion) is needed to change a solid to a liquid;
  d) know why energy (the energy of vaporization) is needed to change a liquid to a gas;
  e) know the definitions of specific energy of fusion and specific energy of vaporization, and be able to perform simple calculations involving specific energies;
  f) be able to explain how something can be cooled by evaporation;
  g) know the effects of pressure and impurities on the melting and boiling points of a substance;
  h) know the definition of linear expansivity, and

be able to use this definition in simple calculations involving expansion;

i) be able to define specific heating energy and be able to perform simple calculations involving specific heating energy.

2 You should know what each of the following is, or what it does:

state of matter   thermometer
bimetallic strip

3 Here are some of the other words that have been used in this chapter. You should know what each word means:

existence   random
vibrate   conversely
modify   evaporate
boil   condense
freeze   solidify
expand

FURTHER QUESTIONS

1 a) Why is a scald from steam usually worse than one from boiling water?

b) Why do vegetables in a pan on a gas ring not cook any quicker if the gas is kept fully on after the water in the pan boils, rather than being turned down?

c) In hot countries, water can be kept cool by keeping it in a porous earthenware pot through which the water can very slowly pass to the outside. How does this keep the water cool?

d) Why should athletes put on a track suit as soon as they have finished running a race?

e) Why is a car cooling system usually pressurized?

f) Why does it take a long time to boil an egg on a high mountain?

g) Why does the jam in the middle of a roly-poly pudding usually seem hotter than the pudding?

2 Reinforced concrete is concrete with iron bars passed through it to strengthen it. Why do you think it is a good thing for the users of reinforced concrete that iron and concrete have the same linear expansivities.

3 The power of the heating element of a kettle is 3 kW.

a) 2 kg of water at a temperature of 0 °C are put in the kettle. How long will it take to boil this water (assuming there are no energy losses)?

b) How long would it take to boil the water if you started with 2 kg of snow at 0 °C instead of 2 kg of water?

(The specific heating energy of water is 4200 J/kg °C; the specific energy of fusion of water is 334 000 J/kg.)

4 A car of mass 1000 kg has four disc brakes, each of mass 5 kg. The material from which the brakes are made has a specific heating energy of 440 J/kg °C. The car is travelling at a speed of 15 m/s when the brakes are applied and it is brought to rest.

a) How much kinetic energy does the car have before the brakes are applied?

b) Assuming all this energy goes into the brakes, what is their temperature rise?

10

LIQUIDS AND GASES

Fig. 10.1   Why does a ship float?

10.1   Introduction

Why does this ship float? Surely its weight should pull it to the bottom of the sea. There must be some other force acting on the ship, pushing it *up* with the same force as its weight pulling it *down*. If this is so, the ship's weight and this upward pushing force are *balanced* forces, that is, equal in size but pushing in opposite directions. Since the forces are balanced, the ship does not accelerate down (to the sea bed) nor up (into the air!)

Gases also exert forces, for example, think of the air inside a bicycle tyre. How do gases exert forces? What rules and laws are there about the way gases behave?

You should know that the particles in liquids and gases are moving around in random directions, and you should know why scientists think that the particles are moving around. But how fast are they moving? Do they all move at the same speed? Do they often bump into each other? It is questions like these which are the concern of this chapter.

10.2   You should know . . .

In order to understand this chapter, you need to know:
a) that forces are measured in newtons;
b) how to measure volumes of solids and liquids;
c) that the density of a substance is the mass of a unit volume, and that its units are kg/m^3;
d) that pressure is the force acting on every square metre, and that its units are pascals (Pa) (the units N/m^2 are sometimes used instead);
e) some of the evidence that makes scientists believe that liquids and gases consist of small particles moving around at random;
f) the relation between the pressure and volume of a gas (Boyle's law);
g) the relation between the pressure and temperature of a gas;
h) the ideas of absolute zero and the kelvin temperature scale.

There is a reminder of some of these ideas later in this chapter.

10.3   Investigating upthrust

Upthrust is the name we give the force which pushes the ship up. Is there always an upthrust on something in water, or does it only happen with ships and other things which float?

Weigh a regular-shaped piece of metal, such as aluminium, with a spring balance as in Fig. 10.2. Make a note of the weight. Submerge the metal in a beaker of water and weigh it again (Fig. 10.3). Make a note of the new reading on the spring balance. Is this reading the same as the first one?

Fig. 10.2
Weighing
aluminium?

Fig. 10.3   Does aluminium lose any
weight when immersed in
water?

If the second reading is less than the first one, it seems unlikely to be because the aluminium really does weigh less; the Earth is surely pulling down on it with the same force as before. The water could be pushing up on the aluminium (for some reason) with a force, or upthrust, which partly balances the weight. To investigate this suggestion further, try the same experiment as before with a *bigger* piece of aluminium. Is there the same upthrust as on the smaller piece when the aluminium is submerged in the water, or more upthrust, or less?

When you submerge the metal in water, it 'pushes to one side', or **displaces**, some water. Find out the *weight* of the water displaced by filling a can (like the one in Fig. 10.4) with water up to the spout, then carefully placing your piece of aluminium in the can. Catch the displaced water coming from the spout in a beaker and find its weight.

Fig. 10.4   How much water does the
aluminium displace?

When Richard did this he found that the weight of water displaced was the same as the loss in weight of his piece of aluminium.

Repeat this experiment for other objects and other liquids. Do you come to the same conclusion?

This idea was first discovered by a Greek called Archimedes over 2000 years ago. Look up the story of Archimedes and try to find out what made him discover this rule.

Copy out and complete the following sentence about **Archimedes' principle**:

'When a body is wholly or partly immersed in a fluid, it experiences an _____ equal to the _____ of the fluid displaced.'

10.4 Why is there an upthrust?

How are you going to explain this upthrust that you have noticed? Fig. 10.5 gives a clue. It is an old oil can with three holes punched in the side. The can is filled with water, which obviously comes out of the holes. There is clearly more pressure behind the bottom jet of water than the other two, because the pressure in a liquid increases the further down you go. This demonstration also reminds you that pressures in liquids act in all directions, so the pressure is able to push the liquid out sideways.

Fig. 10.5 Why does water coming from the bottom hole go further?

The piece of aluminium submerged in water in Fig. 10.3 had water pressure acting on top of it, but had a *greater* water pressure pushing up on the bottom of it, as in Fig. 10.6. It begins to look as if this might explain upthrust.

Water pressure pushes down here

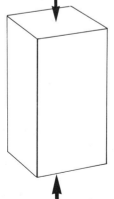

A greater water pressure pushes up here

Fig. 10.6 Calculating the pressure acting on a block of aluminium

To explore this idea a little more fully you need to remember how to *measure* pressure. Pressure is the force acting on each square metre of a surface.

Imagine a convenient rectangular column of any liquid (not necessarily water) such as is illustrated in Fig. 10.7. Here is how to calculate the pressure caused by that liquid on the bottom of the column.

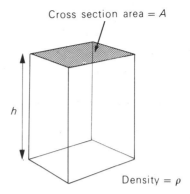

Cross section area = A

h

Density = ρ

Fig. 10.7

Looking at Fig. 10.7,

the *volume* of liquid is $A \times h$ (Why?)

If the density (i.e. the mass for every unit of volume) of the liquid is ρ, then

the *mass* of liquid is $A \times h \times \rho$

If the gravitational field strength is g, then

the *weight* of liquid is $A \times h \times \rho \times g$

This is the force pushing down on the base of the column. The force pushing on each square metre of the base is simply the total force we have just calculated divided by the area of the base:

$$\text{force per unit area} = \frac{A \times h \times \rho \times g}{A}$$

$$= \rho \times g \times h$$

But force per unit area is what we call *pressure*. So the simple formula for calculating the pressure at any depth in a liquid is

$$\text{pressure} = \text{density} \times g \times \text{depth}$$

or

$$P = \rho \times g \times h$$

Now return to the piece of aluminium in water. The aluminium in Fig. 10.3 was 10 cm long. Suppose it is only *just* submerged. There will be no water pressure above it. The water pressure acting on the bottom of the aluminium, 10 cm below the surface, can be calculated using the formula you have just met.

$$P = \rho \times g \times h$$

In this case,

$\rho = 1000 \, \text{kg/m}^3$ (Density of water.)
$g = 10 \, \text{N/kg}$ (Earth's gravitational field.)
$h = 0.1 \, \text{m}$ (depth must be in metres.)

$$P = 1000 \, \text{kg/m}^3 \times 10 \, \text{N/kg} \times 0.1 \, \text{m}$$
$$= 1000 \, \text{Pa}$$

What is the force on the base of the aluminium? The area of the base is

$$0.02 \, \text{m} \times 0.02 \, \text{m} = 0.0004 \, \text{m}^2$$

therefore the force pushing up on the base is

$$1000 \, \text{Pa} \times 0.0004 \, \text{m} = 0.4 \, \text{N}$$

This is exactly the same as the upthrust Richard obtained when weighing his aluminium. So we can explain the upthrust which occurs by the difference in pressures between the top and the bottom of the object in the liquid.

For this section we assumed that the aluminium was *just* submerged. Now it is your turn to do a calculation to show that it does not matter to what depth the aluminium is submerged—the upthrust can always be explained by the *difference* in pressures between the top and bottom surfaces of the aluminium. You could also try to show that it does not matter which way up the aluminium is in the water, the answer is always the same.

10.5   Floating

Now you can understand better why the ship at the start of this chapter floated. It displaced some water, and the weight of the water it displaced was equal to the weight of the ship, so the upthrust was equal to the weight of the ship. This is sometimes called the **principle of flotation**—something which is floating in a liquid displaces its own weight of liquid.

If more cargo is loaded onto the ship, the ship will be heavier, so it will have to displace more water in order to float; in other words, it will sink a little further into the water. On the side of all merchant ships is a sign called the **Plimsoll mark**; an example is illustrated in Fig. 10.8. It is named after Samuel Plimsoll, MP, who was responsible for the Merchant Shipping Act of 1876 which forbade overloaded ships to leave port.

Fig. 10.8   A Plimsol mark

Ships must not be loaded so that they sink below the appropriate load-line. The top line in Fig. 10.8, marked with an 'F', is the fresh-water load line, while the other two are the salt-water load lines for summer and winter. Notice that a ship will sink lower in fresh water since fresh water is less dense than sea water, so the ship has to displace more fresh water than sea water to displace its own weight of water.

The load lines illustrated in Fig. 10.8 were photographed on a small coastal freighter. Ships which are likely to go to other parts of the world have additional load lines. Try to find out what they are.

What do you think the letters 'LR' stand for?

Ice is slightly less dense than water. Most of the iceberg in Fig. 10.9 has to be submerged in order to displace its own weight of water. This makes it a particular hazard for ships; 90% of it is below the surface and therefore invisible.

Fig. 10.9   Most of this iceberg is under water

What do you think is the density of ice compared to water if 90% of an iceberg is below the surface?

10.6   The hydrometer

In the last section you saw that a ship will sink lower in fresh water than in salt water because fresh water is less dense than salt water. The less dense a liquid is, the lower something will sink in it before it displaces its own weight of the liquid.

Fig. 10.10
A wine hydrometer in use

This fact is made use of in the **hydrometer**. This is a simple instrument for measuring the density of liquids, such as the acid in car batteries, milk, beer and wine. The hydrometer in Fig. 10.10 is measuring the relative density (that is, the density compared to water) of some fermenting wine. There is a weight in the bottom of the hydrometer to keep it upright while it is floating. The stem with the figures on it is narrow so that small changes in the density of the liquid produce a large change in the level at which the hydrometer floats.

The reading on the hydrometer in the photograph is 1.020; the density of the liquid is 1020 kg/m³. There is still some sugar in solution to be fermented.

10.7 Balloons

Fig. 10.11

This hot-air balloon is floating. It is another example of Archimedes' principle in action. Liquids and gases together are called **fluids**, and Archimedes' principle applies to all fluids, not just liquids.

The hot-air balloon is displacing some air. Since the balloon is floating at a steady height, the weight of air displaced must be equal to the weight of the balloon (including the passengers and their basket). The balloon is filled with hot air to make it sufficiently light to float; hot air is less dense than cold air.

Other balloons are made light by using a much lighter gas than air. Hydrogen is the lightest gas and was used for the big airships of the 1930s. Unfortunately, hydrogen is highly flammable, and airships were abandoned after some horrifying fires.

Fig. 10.12   A hydrogen airship

The balloon in Fig. 10.13 carries meteorological instruments high into the atmosphere. It is filled with hydrogen. The weight of air it displaces is much greater than its own weight, so the upthrust is greater than its own weight. When it is released it will go high into the atmosphere and send back information about the upper atmosphere to meteorologists.

Fig. 10.13   This balloon will carry meteorological instruments high into the atmosphere

Helium is a safer gas to use, but it is four times as dense as hydrogen and very expensive.

10.8 A numerical example

A hot-air balloon is tied to the ground on a windless day. The balloon contains $1500\,\text{m}^3$ of hot air. The density of this hot air is $0.8\,\text{kg/m}^3$. The mass of the balloon (not including the hot air) is $500\,\text{kg}$. The density of the surrounding air is $1.3\,\text{kg/m}^3$.

a) What is the tension in the rope holding the balloon to the ground?

b) If the balloon is released, what is its initial acceleration?

Part a)

The upthrust on the balloon is greater than its weight. Since the balloon does not move, the tension in the rope must be equal to the difference between the upthrust on the balloon and the weight of the balloon.

To calculate the weight of the balloon:

$$\text{mass of hot air} = \text{density} \times \text{volume}$$
$$= 0.8\,\text{kg/m}^3 \times 1500\,\text{m}^3$$
$$= 1200\,\text{kg}$$

mass of balloon = $500\,\text{kg}$ (from question)
total mass of balloon + hot air = $1700\,\text{kg}$

$$\text{total weight of balloon} = mg$$
$$= 1700\,\text{kg} \times 10\,\text{N/kg}$$
$$= 17\,000\,\text{N}$$

To calculate upthrust on the balloon:

$$\text{upthrust} = \text{weight of air displaced}$$
$$\text{mass of air displaced} = \text{density} \times \text{volume}$$
$$= 1.3\,\text{kg/m}^3 \times 1500\,\text{m}^3$$
$$= 1950\,\text{kg}$$

$$\text{weight of air displaced} = mg$$
$$= 1950\,\text{kg} \times 10\,\text{N/kg}$$
$$= 19\,500\,\text{N}$$

If the balloon is to remain still, the total force down must be equal to the total force up. An additional force of

$$19\,500\,\text{N} - 17\,000\,\text{N} = 2500\,\text{N}$$

is needed for the forces to balance. The tension in the rope is $2500\,\text{N}$.

Part b)

If the balloon is released, there will be an unbalanced force of $2500\,\text{N}$ acting on the balloon causing it to accelerate upwards.

Using
$$F = ma$$
$$2500\,\text{N} = 1700\,\text{kg} \times a$$

(Where has the $1700\,\text{kg}$ come from?)

$$a = \frac{2500\,\text{N}}{1700\,\text{kg}}$$
$$= 1.47\,\text{m/s}^2$$

The balloon accelerates upwards at $1.47\,\text{m/s}^2$ (assuming air resistance can be ignored).

10.9 Hydraulic machines

Fig. 10.14   This digger is operated by hydraulics

The various parts of the digger in Fig. 10.14 are moved by a liquid under pressure pushing pistons. Machines like this are called **hydraulic** machines. They depend on the fact that liquids are almost incompressible, and that the pressure in a liquid is transmitted equally to all parts of the system. Corners and bends in the pipes have no effect on the pressure in a hydraulic system. Some of the hydraulic pipes on the digger are flexible so that they can move with the moving arms of the digger. It would be virtually impossible to construct a digger like this using a system of levers to transmit the forces to where they are needed.

A simpler hydraulic machine is the hydraulic jack Michael is using in the next photograph. How does the small force Michael applies become a force large enough to lift the vehicle? Look at Fig. 10.16.

Fig. 10.15   Another example of hydraulics in action

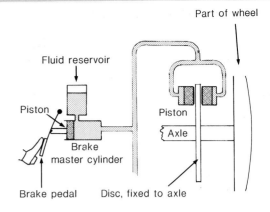

Fig. 10.17   The hydraulic brake system of a car

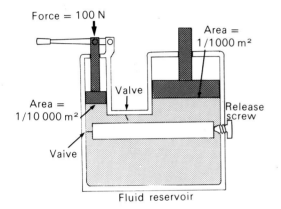

Fig. 10.16   The hydraulic jack

You can see that Michael pushes down on a piston of *small* cross-sectional area, and the vehicle rests on a piston of *large* cross-sectional area. Suppose the cross-sectional area of the small piston was $1/10\,000\,\text{m}^2$ and that Michael pushes down so that the force on the small piston is 100 N.

The pressure on the small piston will be

$$\frac{100\,\text{N}}{1/10\,000\,\text{m}^2} = 1\,000\,000\,\text{Pa}$$

Suppose the large piston has a cross-sectional area of $1/1000\,\text{m}^2$. The force acting on this piston will be

$$1\,000\,000\,\text{Pa} \times 1/1000\,\text{m}^2 = 1000\,\text{N}$$

This is 10 times greater than the force on the small piston.

What is the mechanical advantage of this system?

How far up will the large piston go for every 10 cm that the small piston goes down?

Car brakes are usually worked by a hydraulic system. Fig. 10.17 shows a simplified diagram of part of a car braking system. Look carefully at the diagram and try to decide how the system works.

10.10   Some points concerning gases

You may already have investigated gases in previous years. Here are some quick reminders of some of the ideas you may have met.

1   An important piece of evidence for gases consisting of moving particles is **Brownian motion**. A suitable apparatus for observing Brownian motion is illustrated in Fig. 10.18. Smoke (for example, from a burning straw) is put into the smoke cell, illuminated by means of a lamp and viewed under a microscope. The smoke appears as small pin-pricks of light moving around in random directions and constantly changing direction (Fig. 10.19). Smaller, faster, invisible air particles bombard the smoke particle causing it to move in this fashion.

Fig. 10.18   Apparatus for examining Brownian motion

Path of smoke particle

Fig. 10.19   Typical path of a smoke particle in the apparatus in Fig. 10.18

2 The relation between the pressure and volume of a gas can be investigated using apparatus such as that illustrated in Fig. 10.20. Pressure is exerted on the air in the tube using a car tyre pump. A typical graph of pressure against volume is shown in Fig. 10.21; as the pressure increases the volume decreases. If you look carefully at the graph you can see that as the pressure doubles, the volume halves. We say the pressure is *inversely proportional to* the volume. This is known as **Boyle's law**.

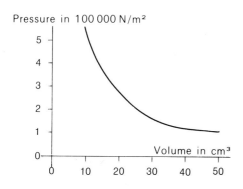

Fig. 10.20 Investigating Boyle's law

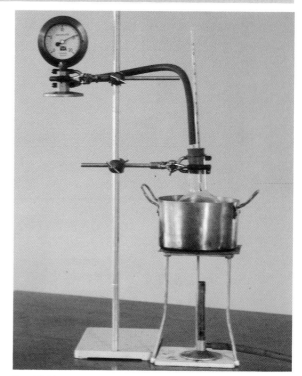

Fig. 10.22 Apparatus to investigate how the pressure of a gas depends on its temperature

Fig. 10.21 A graph of pressure against volume for a gas

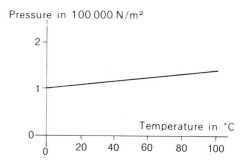

Fig. 10.23 A graph of pressure against temperature for a gas

3 A suitable apparatus for investigating the relation between the pressure and the temperature of a gas is shown in the next photograph. Air in the flask is heated by the water. The pressure in the flask is measured at a number of temperatures in order to obtain data for a graph like the one in Fig. 10.23. Although this is a straight line graph, it does not go through the origin and so the pressure is not proportional to the temperature in degrees Celsius.

4 The graph shown in Fig. 10.23 can be *extrapolated* back to the temperature axis (Fig. 10.24) to find the temperature at which the pressure of the gas would be zero, assuming it remains a gas (which it will not in practice). This temperature turns out to be $-273\,°C$ for all gases. It is called **absolute zero**; scientists believe that this is the lowest temperature that can be obtained. We can define a new temperature scale, called the **kelvin scale of temperature**, which has $0\,K$ at absolute zero. On this scale, water freezes at $273\,K$ and boils at $373\,K$. A temperature *interval* of $1\,K$ is the same size as a temperature interval of $1\,°C$.

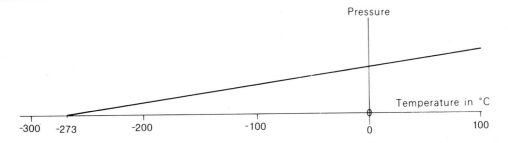

Fig. 10.24   Finding absolute zero

10.11   The general gas law

The last section reminded you of two *facts* about gases.

1 Boyle's law. The pressure of a gas is inversely proportional to its volume, provided the temperature remains constant. Therefore a graph of pressure against (1/volume) is a straight line passing through the origin (Fig. 10.25). The *gradient* of this graph is

$$\frac{P}{1/V}$$

which is equal to $P \times V$. Since it is a straight line graph, the gradient is the same at all points on the graph, and we can write

$$P \times V = k$$

k = constant = gradient of graph.

Pressure in 10^5 N/m²

5 —
4 —
3 —
2 —
1 —
0 —
    0   0.02   0.04   0.06   0.08   0.10
                1/Volume in 1/cm

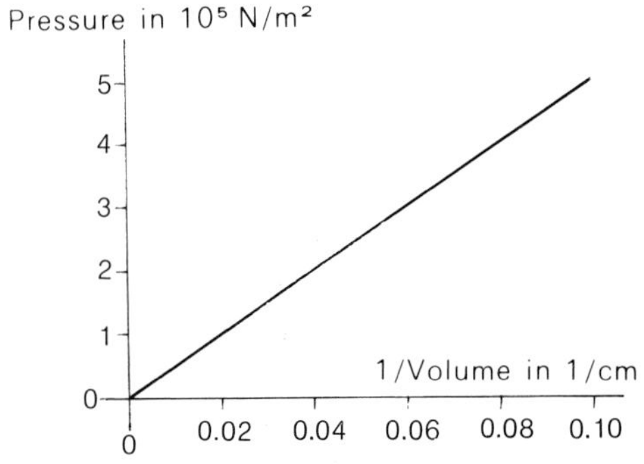

Fig. 10.25   A graph of P against $1/V$ for a gas

2 The pressure of a gas is proportional to its temperature, provided the temperature is measured in kelvin. A graph of P/T is a straight line, so the gradient is the same at all points on the graph. We can write

$$\frac{P}{T} = C$$

C = constant = gradient of graph.

These are two of the **gas laws**. They are a *summary* of the way gases behave under certain conditions. (This is what a law does in science; it tells you what happens in particular circumstances. It does not offer an explanation as to *why* something happens.)

These two laws can be combined together so that we can calculate what happens, for example, to the volume of a gas if the pressure and temperature both change together, as might happen in a chemical reaction.

Combining the two equations above gives

$$\frac{P \times V}{T} = \text{constant}$$

This is the **universal gas equation**. It is telling you that, for a given mass of gas,

$$\frac{\text{pressure} \times \text{volume}}{\text{temperature}}$$

always remains the same.

We can check that this equation agrees with what we have had before. If the temperature does not change, then this reduces to $P \times V$ is constant, which is Boyle's law. If the volume does not change, then the equation reduces to P/T is constant, which was the second equation at the start of this section. The equation also predicts that if the pressure does not change, then V/T remains constant. This is known as **Charles' law**.

Using the apparatus illustrated in Fig. 10.26, and anything else you think you need, design an experiment to test Charles' law. Remember that the temperature must be measured in kelvin.

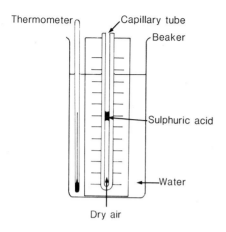

Fig. 10.26   How could you use this apparatus to investigate the way the volume of a gas depends on its temperature?

Another way of writing the universal gas equation is as follows. Suppose the pressure, volume and temperature of a sample of gas at a certain time are P_1, V_1 and T_1; some time later the values are P_2, V_2 and T_2; we can write

$$\frac{P_1 \times V_1}{T_1} = \frac{P_2 \times V_2}{T_2}$$

10.12   A numerical example using the gas laws

A meteorological balloon like the one in Fig. 10.13 contains $20\,m^3$ of hydrogen at atmospheric pressure $(10^5\,N/m^2)$ and a temperature of $15\,°C$. What will be the volume of the hydrogen when the balloon has reached $10\,000\,m$ (about the height of Everest) where the pressure is one-third that of the pressure at sea level and the temperature is $-20\,°C$? Assume the balloon fabric can expand sufficiently to accommodate the gas.

DATA

	Before	After
volume	$20\,m^3$	V_2
pressure	$10^5\,N/m^2$	$3.3 \times 10^4\,N/m^2$
temperature	$288\,K$	$253\,K$ (NB kelvin)

Using

$$\frac{P_1 V_1}{T_1} = \frac{P_2 V_2}{T_2}$$

$$\frac{10^5\,N/m^2 \times 20\,m^3}{288\,K} = \frac{3.3 \times 10^4\,N/m^2 \times V_2}{253\,K}$$

$$V_2 = \frac{253 \times 10^5 \times 20}{288 \times 3.3 \times 10^4}\,m^3$$

$$= 53\,m^3$$

10.13   How fast does a molecule move?

You know by now that the air molecules all around you are moving at high speeds, but what do we mean by high speeds—$10\,m/s$, $100\,m/s$, $1000\,m/s$?

To answer this question we need to think about why a gas exerts a pressure at all. Look at Fig. 10.27. Marbles are being poured onto a lever arm balance and as they bounce off the balance they change their momentum. You learned in Chapter 5 that an impulse (force×time) is needed in order to change momentum. The force between the marbles and the balance pan is actually being measured.

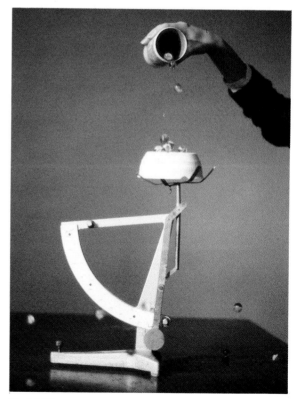

Fig. 10.27

Gases exert a pressure for the same reason. Whenever a molecule hits a surface it bounces off and so changes its momentum. All the little impulses needed for each one of these momentum changes add up to a large impulse. Let us now use this idea to calculate how fast a gas molecule might be moving in order that all the little impulses can give rise to the pressure we observe.

Imagine a box like the one in Fig. 10.28. In it is a molecule moving with velocity v along the box in the direction of the arrow. The molecule has a mass m, so its momentum is mv.

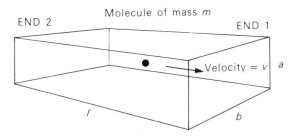

Fig. 10.28

When it gets to the end of the box labelled 'END 1' the molecule rebounds. Suppose it has a perfectly elastic collision, so it does not lose any energy and it rebounds with velocity $-v$ (remember velocity is a vector quantity); its momentum is now $-mv$ and the *change* in the molecule's momentum is $2mv$.

The molecule will then go to 'END 2', bounce off and return to END 1, bounce off (changing its momentum by $2mv$ again), and so on.

Since it is travelling with a speed v all the time, it will travel a total distance of v metres in 1 second. This distance is made up of many little 'journeys' up and down the box. The length of one 'journey' is $2l$ (Why?) so the molecule will make $v/2l$ journeys in 1 second. Therefore it will hit END 1 $v/2l$ times in 1 second.

Each time it hits this end, it changes its momentum by $2mv$, so the *total* change in momentum of this molecule at END 1 in 1 second is

$$\frac{2mv \times v}{2l} = \frac{mv^2}{l}$$

Since impulse = change in momentum, the impulse at END 1 due to this one molecule is mv^2/l

$$Ft = \frac{mv^2}{l}$$

In this case, $t = 1$ second (Why?), so the force on END 1 due to this one molecule is mv^2/l.

Of course, no real box will have just one molecule in it! Suppose there are N molecules in the box. They will not all be moving along the box in the same direction as the one illustrated. Some will move up and down, others will move from side to side. Let us assume, on average, one-third of the molecules in any cube are moving between ENDS 1 and 2. (Do you think this is reasonable?) So the force on END 1 due to $N/3$ molecules is

$$\frac{1}{3} \frac{Nmv^2}{l}$$

The *pressure* on END 1 is the force/area. Therefore,

$$\text{pressure} = \frac{1}{3} \frac{Nmv^2}{abl}$$

But abl is the volume V of the box, so

$$\text{pressure} = \frac{1}{3} \frac{Nmv^2}{V}$$

Here, then, is an equation which relates the *pressure* of a gas (which we can measure) to the *speed* of the molecules (which we cannot easily measure). The equation can be simplified a little. Nm is the *total* mass M of *all* the molecules in the box, so

$$P = \frac{1}{3} \frac{Mv^2}{V}$$

M/V is the density ρ of the gas (mass per unit volume), so

$$P = \tfrac{1}{3}\rho v^2$$

The density of the air in your home is $1.2 \, \text{kg/m}^3$ and the pressure is about $10^5 \, \text{Pa}$. Putting these figures into the equation above gives

$$10^5 \, \text{Pa} = \tfrac{1}{3} \times 1.2 \, \text{kg/m}^3 \times v^2$$

$$v^2 = \frac{3 \times 10^5 \, \text{Pa}}{1.2 \, \text{kg/m}^3}$$

$$v = 500 \, \text{m/s}$$

This section has predicted that the molecules around you are moving at a speed of $500 \, \text{m/s}$. At this rate, a molecule would travel the $300 \, \text{km}$ from London to Manchester in 10 minutes.

10.14   Different velocities

In fact, not all air molecules travel at a speed of $500 \, \text{m/s}$. Just as with liquids (see Section 9.8), there is a *range* of speeds, as shown in Fig. 10.29. Some molecules, as a result of chance collisions, are moving very slowly or have even stopped; others are moving very fast; most are moving at a speed between these two extremes. The speed v in the equation $P = \frac{1}{3}\rho v^2$ must represent some average speed of the molecules.

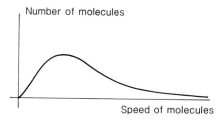

Fig. 10.29 Particles in a gas are found with a range of velocities

The same equation tells you that the speed of the molecules depends on the *density* of the gas. At a given pressure, the *larger* the density the *smaller* the speed. Suppose you have two gas jars, one full of oxygen molecules and the other full of hydrogen molecules, both at the same pressure. Which gas is less dense—oxygen or hydrogen? Which gas has, on average, the faster moving molecules?

In Chapter 11 there is described an experiment to measure how fast sound waves travel through air. The sound waves are carried by the molecules in the air; the faster the molecules are moving, the faster the sound waves can travel. In which gas do you think sound travels faster—oxygen or hydrogen? When you reach Chapter 11 and measure the speed of sound, try to invent a way of adapting the experiment so that you can test whether or not your answer to the last question was correct.

10.15 Diffusion

Fig. 10.30 will remind you of something you have seen before—brown bromine gas slowly diffusing through air in a gas jar. In Section 10.13 we found that molecules are moving very fast. Why does the bromine diffuse so slowly?

(Bromine is denser than air—7.2 kg/m^3 instead of 1.2 kg/m^3—so will the molecules of bromine be moving faster or slower than molecules of oxygen or nitrogen? Use the equation $P = \frac{1}{3}\rho v^2$ to calculate the speed of bromine molecules at atmospheric pressure, which is $100\,000 \text{ Pa}$.)

Fig. 10.30
Why does the bromine take a long time to diffuse into the air?

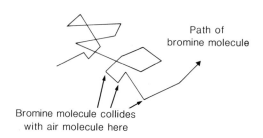

Fig. 10.31 Typical path of a diffusing bromine molecule

As you know, the reason for the slow progress of the bromine up the gas jar is that the air molecules in the gas jar impede the progress of the bromine molecules. Bromine molecules do not travel straight up the gas jar but keep bumping into air molecules and become deflected from their path, as illustrated in Fig. 10.31.

You can understand this slow progress a little better by playing the following game. You need a sheet of isometric graph paper as illustrated in Fig. 10.32, a die and a pencil. Imagine a bromine molecule at the centre of this piece of paper. It could be travelling in *any* direction, but to keep things simple imagine it can only go along one of the six lines leading from the point where the molecule is. Throw the die to decide in which direction the molecule goes (each direction is numbered at the top of Fig. 10.32). Draw a line along this direction to the next point where lines meet. Imagine that your bromine molecule collides with an air molecule here. Throw your die again to determine in which direction your bromine molecule now goes. Carry on for, say, 25 'moves'. Does your bromine molecule end up very far from the starting point? Can you see why diffusion is such a slow process?

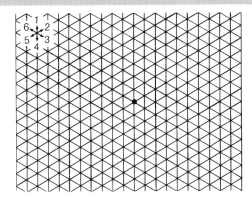

Fig. 10.32   Isometric graph paper

Fig. 10.33   A 'random walk' in progress

The path that your bromine molecule follows is known as a **random walk** (or sometimes a 'drunkard's walk').

Of course, this **simulation** is a big oversimplification of the real thing. The distance the bromine goes between each collision will not be the same because the air molecules are not arranged in a neat pattern like Fig. 10.32. In what other ways do you think this simulation is an oversimplification?

This is the sort of simulation that a computer is good at doing. Fig. 10.34 shows a computer-generated random walk. A computer can be more sophisticated than the simple model using isometric graph paper. For example, you could have many more directions and a varying distance between collisions. If you know how to write computer programs, you might like to try to write a random walk program.

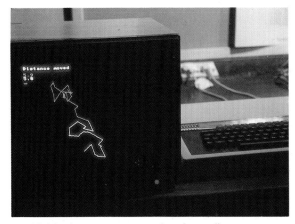

Fig. 10.34   A computer generated random walk

SUMMARY

Now that you have finished studying this chapter on liquids and gases, there are a number of things you should know and be able to do.

1 You should:
   a) know Archimedes' principle and be able to perform simple calculations using this principle;
   b) know that pressure in a liquid acts equally in all directions;
   c) know and be able to apply the principle of flotation;
   d) know what a hydrometer is and be able to explain how it works;
   e) be able to explain how a hydraulic jack works;
   f) understand how hydraulic brakes on a car work;
   g) know some of the evidence that convinces scientists that matter is made up of particles;
   h) know Boyle's law;
   i) know Charles' law;
   j) know the universal gas law;
   k) be able to apply the above laws in simple calculations involving gases;
   l) understand what is meant by 'inverse proportionality';
   m) understand the kelvin temperature scale;
   n) know roughly how fast a molecule moves;
   o) understand how gases diffuse.

2 You should know what each of the following is, or what it does:

   upthrust   piston
   simulation   random walk
   extrapolation

3 These are some of the other words that have been used in this chapter. You should know what each word means:

submerge   displace
float   transmit

illuminate   bombard
interval   represent
determine   oversimplification
deflect   sophisticated

FURTHER QUESTIONS

Assume g = 10 N/kg.

1 Explain the following.
  a) A ship rides higher in salt water than it does in fresh water.
  b) As a swimmer is pulled into a boat, the swimmer appears to become heavier.
  c) An egg sinks in fresh water, but floats in salt water.
  d) A boat sinks lower into the water as cargo is loaded into it.

2 A buoy of mass 300 kg and volume 1 m³ is anchored in water so that two-thirds of it is below the surface. The density of water is 1000 kg/m³.
  a) What volume of water is displaced by the buoy?
  b) What weight of sea water is displaced?
  c) What is the upthrust on the buoy?
  d) What is the weight of the buoy?
  e) What is the tension in the cable anchoring the buoy to the sea bed?

3 Fig. 10.35 shows a diagram of a simple hydraulic press. The small piston has a cross-sectional area of 4 cm² and the large piston has a cross-sectional area of 80 cm².

a) What is the minimum force that can be applied to the small piston to just lift a load of 200 N on the large piston?
b) Why is this the *minimum* force?

4 Which of the following would change the rate of diffusion of bromine through air? Explain your answers and state whether, if a change does occur, the diffusion would become slower or faster.
  a) Increase the temperature of the apparatus.
  b) Reduce the pressure of the air through which the bromine is diffusing.
  c) Increase the quantity of bromine.

5 Before a car starts on a journey, the pressure in the tyres is 2×10^5 Pa and the temperature of the air is 20 °C. After the car has been travelling for some time, the temperature of the air in the tyres has increased to 30 °C.
  a) Why has the temperature of the air increased?
  b) What will be the new pressure of the air in the car tyres? (Assume the volume remains constant.)
  c) Explain why the pressure of the air in car tyres should be checked *before* the start of a journey.

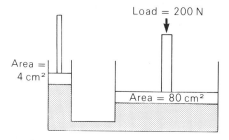

Fig. 10.35

11

OSCILLATIONS AND SOUND

Fig. 11.1   What keeps this clock to time?

11.1   Introduction

The regular backwards and forwards motion of this pendulum is responsible for this clock keeping the right time. You may well have done some experiments with pendula before (see, for example, *Physics in Action* Chapter 4).

We call this backwards and forwards motion an **oscillation**. Oscillations are extremely important to engineers and physicists. They are responsible for many things other than making a clock keep time. In this chapter you will learn, for example, that oscillations make music and destroy bridges. You will learn that something has to oscillate in order that waves can be made.

11.2   You should know . . .

You can understand most of this chapter without knowing much other physics. However, to understand all of it fully, you should:

1 understand Newton's first and second laws of motion, particularly that acceleration is proportional to the unbalanced force;
2 know that the gradient of a distance-time graph is the velocity, and that the gradient of a velocity-time graph is the acceleration;
3 know the difference between a longitudinal (compression) wave and a transverse wave;
4 have used a Slinky spring to investigate the simple properties of waves.

11.3   Three simple oscillators

Fig. 11.2   A simple pendulum

When Sheila releases the pendulum bob (Fig. 11.2) it will swing to and fro. The number of to and fro movements it makes in 1 second is called the **frequency** of the oscillation. Frequency is measured in **hertz** (Hz); 1 Hz is one oscillation per second.

Another word, the meaning of which you must know, is the **period** of an oscillation. This is the time taken for each oscillation to occur. If, for example, the *frequency* of an oscillation is 3 Hz, there will be three oscillations per second. The time taken for each one of those oscillations will be $\frac{1}{3}$ second, so the *period* of the oscillation is $\frac{1}{3}$ second. You should be able to see that the period of an oscillation is 1/frequency of oscillation.

Assemble a simple pendulum for yourself, as in the photograph. Make the length of the pendulum about 0.5 m. Pull the bob (the mass on the end of the string) back about 10 cm and then release it so that the pendulum swings. The **amplitude** of this oscillation is 10 cm; that is, the distance from the middle of the swing, called the **mean position**, to the furthest point to which it swings, is 10 cm (see Fig. 11.3).

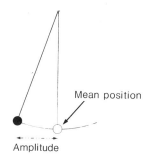

Fig. 11.3   The amplitude of an oscillation

Does the amplitude stay the same all the time?

How can you alter the *frequency* of the oscillation? (For example, does the amplitude, or the length of the pendulum, or anything else, alter the frequency?)

Figs 11.4 and 11.5 show two other simple oscillators. Set these up for yourself and investigate how to change the frequency of these oscillators. The mass hanging on the end of the spring should be between 100 g and 500 g. The mass should be pulled down a short distance and then released, so that it oscillates up and down. You may be surprised at one of the effects which occurs with this oscillator.

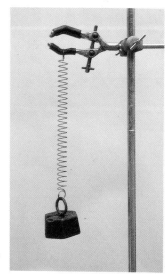

Fig. 11.4
A simple oscillator to investigate

Fig. 11.5
This trolley will oscillate
along the bench
between the springs

Fig. 11.6　The force on an oscillating body acts towards the
centre of the oscillation

With each of the three oscillators you have looked
at, at what point does it move with the maximum
speed? At what point does it move with zero speed?
Where does it have maximum acceleration?

The first two of these questions should be easy
enough to answer, but the third will need a little
more thinking about. It is probably easiest to think
about the trolley tethered between two springs
(Fig. 11.5) to decide where the maximum acceleration is. This will occur when there is the greatest
unbalanced force acting on the trolley—why? Will
there be the greatest force when the trolley is
in the middle, or at the end, or where? Give a
reason for your answer. Where, then, is the greatest
acceleration?

All these oscillating motions are examples of
simple harmonic motion. In all cases the oscillations
are **damped**; i.e. the amplitude gradually decreases
and the oscillations 'die down'. What causes this?
Where does the energy go?

In these examples, notice that the force acting on
the body is always pulling it backwards towards the
centre of the oscillation. Also, the force is *proportional
to* the displacement (the distance from the centre).
This is most obvious in the case of the trolley
between the springs; if the springs obey Hooke's law
then the tension in the springs is proportional to
the extension of the springs. These are the *conditions*
for simple harmonic motion to occur. A formal
definition of simple harmonic motion would be:

> *Simple harmonic motion occurs when the force
> acting on a body is proportional to the displace-
> ment of that body from a fixed point, and is in the
> opposite direction to the body's displacement.*

Look at Fig. 11.7. In which of these examples do
you think oscillations might occur? In which cases
do you think the oscillations might well be simple
harmonic?

You have been introduced to many new terms in
this section. To make sure you understand what
they all mean, copy out the following list and write
a short sentence, or draw a small diagram, to
explain what each item in the list means.

frequency　period　hertz　amplitude
mean position　displacement　damped

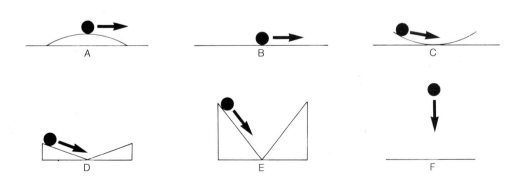

Fig. 11.7　How many of these oscillations could be simple
harmonic?

11.4 A displacement-time graph for simple harmonic motion

In Fig. 11.8 Sheila and Richard are trying to obtain a graph of displacement against time for a large pendulum which has been made from a broomstick handle with two bricks tied to the bottom. There is a brush dipped in ink at the bottom of the pendulum and this draws a graph on the sheet of paper that Sheila is pulling at a steady speed beneath the pendulum.

Fig. 11.8
Trying to obtain a displacement-time graph for a simple pendulum

Do you think it is easy to pull the paper at a steady rate? Can you think of a better way of pulling the paper? Can you think of any other improvements to the experiment?

Assemble the apparatus for yourself, either exactly as in the photograph or with any improvements you can think of. Look at the shape of the graph you obtain. It is called a **sine wave**. (If you plot sine x against x you obtain this shape.)

Does your graph show that the oscillation was damped? If so, what is it about the graph that shows you the oscillation was damped?

This experiment is obviously a fairly crude way of obtaining a displacement against time graph. Can you think of any other ways of obtaining such a graph, perhaps with a different oscillator?

It is quite easy to deduce the shape of a *velocity* against time graph for simple harmonic motion from the displacement-time graph. Remember that you can calculate velocity from the *slope* of a displacement-time graph (see Section 2.7). A displacement-time graph for a simple harmonic oscillator is drawn in Fig. 11.9(a), and underneath is the corresponding velocity-time graph. Where the displacement-time graph has zero slope, the velocity is zero; where the slope is steep, the velocity is high, and so on.

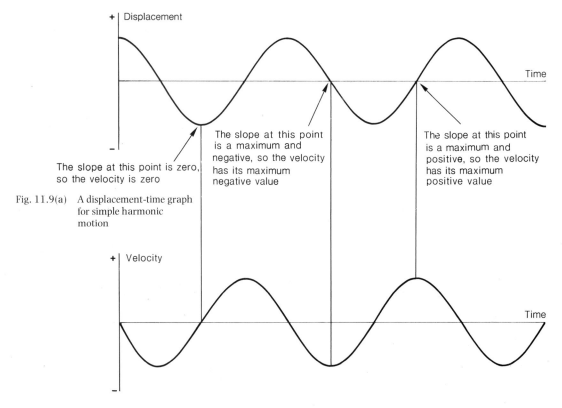

Fig. 11.9(a)   A displacement-time graph for simple harmonic motion

The slope at this point is zero, so the velocity is zero

The slope at this point is a maximum and negative, so the velocity has its maximum negative value

The slope at this point is a maximum and positive, so the velocity has its maximum positive value

Fig. 11.9(b)   A velocity-time graph for simple harmonic motion

Notice that these graphs show that the velocity is *greatest* when the displacement is zero (i.e. the oscillator is at the mid-point), and the velocity is zero when the displacement is greatest. Does this agree with what we decided about simple harmonic motion in the last section?

Draw the two graphs shown in Fig. 11.9, and then, remembering that acceleration can be calculated from the gradient of a velocity-time graph, draw underneath your graphs the acceleration-time graph for simple harmonic motion.

Where does your graph show the maximum acceleration occurs? Does this agree with what you decided in Section 11.3?

11.5   Resonance

The bridge in Fig. 11.10 was certainly not *built* like that! The photograph shows it oscillating with a twisting motion. The bridge was built across the Tacoma Narrows near Seattle, USA, in 1939. As soon as it had been built it was found that it oscillated up and down when the wind blew, and it was christened 'Galloping Gertie'. One windy day in 1940, this twisting motion started; the oscillations built up into a large amplitude which overstressed the bridge with the eventual result shown in Fig. 11.11. (A similar thing happened to the Chain Pier at Brighton in 1836.)

Fig. 11.10   What made this bridge over the Tacoma Narrows oscillate like this and then collapse?

Fig. 11.11

How could the wind cause such destructive oscillations? Notice that it is not a question of the *pressure* of the wind *pushing* the bridge. (This was the cause of the collapse of the Tay bridge one stormy night in December 1879.) The bridge was vibrating up and down, at right angles to the wind direction.

Fig. 11.12   The Tay bridge after being blown down by the wind

To understand how such large oscillations can build up, look at the simple example in Fig. 11.13. To make Nicholas swing higher (i.e. with a larger amplitude), Judith gives him a little push every time he comes near her at the end of each oscillation. Judith pushes him at the *same frequency* as the **natural frequency** of oscillation of the swing, and the amplitude of the swing increases. This is an example of a **resonant system**. A periodic force (from Judith) is applied to an oscillator (the swing) at the same frequency as the natural frequency of the oscillator.

Fig. 11.13   A simple resonant system

Vibrations in a car or lorry provide another example of resonance. As the engine rotates it exerts an oscillating force. If this occurs at the same frequency as the natural frequency of vibration of any part of the vehicle, that part will 'resonate' and build up a large amplitude oscillation.

But how can a steady wind provide an oscillating force? When wind comes to an obstruction it has to go round it. If the obstruction is **streamlined** then the airflow is as illustrated in Fig. 11.14(a). However, an obstruction like the one in Fig. 11.14(b) gives rise to little **eddies**, or **vortices**, alternately one side then the other. This process is called **vortex shedding**. It can be seen taking place in Fig. 11.15 as the air, mixed with smoke to make the flow visible, comes from the left past a flat obstruction.

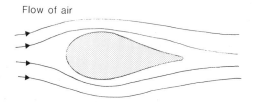

Fig. 11.14(a)   Airflow round a streamlined obstruction

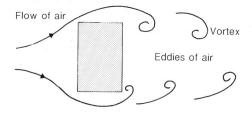

Fig. 11.14(b)   Airflow round a non-streamlined obstruction

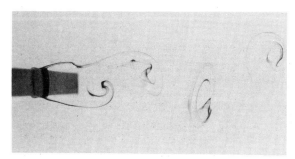

Fig. 11.15   Vortex shedding made visible by smoke

These vortices are at a lower pressure than the rest of the air, so the obstruction is pushed towards each vortex in turn, first one way and then the other.

The frequency of vortex shedding depends on the wind speed. If this frequency happens to be the same as the natural frequency of vibration of the obstruction, resonance will occur and a large amplitude oscillation can build up. This is exactly what happened in the case of the Tacoma Narrows bridge.

Tall chimneys, bridges, cables and aerial masts are all examples of things which may be subject to wind-induced oscillations. There are many complicated ways in which these structures might oscillate, and there are a number of ways in which engineers reduce these oscillations. For example, an obstruction can be streamlined, as in the case of the Humber road bridge. The oscillations can be damped, that is, the energy absorbed in some way. Overhead cables often have special dampers like those in Fig. 11.18 to reduce cable oscillation. The 'straikes' at the top of the chimney in Fig. 11.19 help to prevent vortices forming at all.

Can you think of any other sorts of oscillation (not necessarily wind-induced ones) which might be a nuisance? Do you know how these oscillations might be reduced?

Fig. 11.16   The Humber bridge has a streamlined road deck

Fig. 11.17
This car shock absorber damps the oscillations of the road spring

Fig. 11.18
Dampers on
overhead electricity
cables

Fig. 11.21
Investigating
standing waves on a
string

Fig. 11.19
The 'straikes' on this
chimney help to
destroy vortices and
reduce oscillations of
the chimney

The red object attached to the string near the camera is a vibrator which can be made to vibrate at any frequency by adjusting the frequency of the electrical oscillator which powers it.

If possible, look at this apparatus in action for yourself. If the vibrator is set in motion at a low frequency, say 5 to 10 Hz, the string will vibrate a little. This is called a **forced oscillation** of the string. If the frequency of the oscillator is gradually increased, at one particular frequency the string will build up into a large amplitude oscillation, as in Fig. 11.22. This is called a **standing wave**. Here is another example of resonance: the frequency of oscillation of the vibrator is the same as the natural frequency of the string.

11.6   Vibrations and music

Fig. 11.20   Music is caused by oscillations

Not all oscillations are harmful! Most people enjoy some sort of music, and all musical instruments oscillate in some way. For example, when a guitar string is plucked, or a piano string hit, the string vibrates at its natural frequency and produces a sound wave.

But middle C, for example, sounds quite different when played on a guitar than when played on a piano. Why?

To investigate this problem, you could use apparatus like that illustrated in Fig. 11.21. A 10 N weight is hanging off the end of the string after it passes over the pulley at the far end of the bench.

Fig. 11.22   A standing wave formed using the apparatus in the previous photograph

If the frequency of the vibrator is gradually increased beyond this particular value, the amplitude of the string's vibrations decreases again, and then, at a new particular frequency, a new standing wave pattern appears, as illustrated in Fig. 11.23.

Fig. 11.23   To form this standing wave, the string was vibrated at twice the frequency of the string in the previous photograph

At what frequency does this occur compared to the frequency for the first standing wave? Can you obtain any other standing wave patterns? At what frequencies do you find them?

When the string of, say, a guitar is plucked, it vibrates at its natural frequency and gives out a note of that frequency (or **pitch**). These last experiments show, however, that a string will also vibrate at higher frequencies, so as well as giving out the **fundamental frequency**, other, higher, frequencies, called **overtones**, are also present. The **quality** (or **timbre**) of a musical note depends on the number, frequency and energy of these overtones. If possible, try plucking a guitar string at different places along its length. Can you hear the difference in quality which depends on where you pluck the string? If you pluck it near one end, you tend to get more overtones. Can you suggest why?

Fig. 11.24 shows oscilloscope traces of the note middle C from a tuning fork, piano and guitar. (These sounds were recorded by a VELA, using a microphone, and then played out into the oscilloscope.) Although the fundamental frequency is the same, the overtones, and hence the quality of the note, differ. (There were no overtones from the tuning fork.)

Returning to the apparatus illustrated in Fig. 11.21, suppose a *larger* weight is now hung on the end of the string, so thare is a greater tension in the string. Does the resonant frequency, and hence the string's natural frequency of oscillation, change? Fig. 11.25 shows the tuning mechanism for a guitar, which is for adjusting the string's natural frequency of vibration. How do you think it works? What has it to do with this experiment?

Another change you could make to the above experiment is to use a *shorter* length of string. How does this change the natural frequency? If you play a stringed instrument you will know that you play different notes by 'shortening' a string with your fingers (Fig. 11.26).

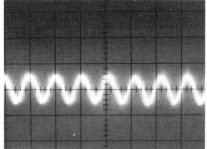

a) Tuning fork

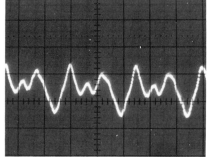

b) Piano

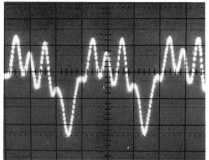

c) Guitar

Fig. 11.24 Oscilloscope traces showing the waveforms of notes from a tuning fork, piano and guitar

Fig. 11.25 Tuning a guitar

Fig. 11.26 Different notes are played on a guitar by changing the effective length of the string, and so changing the frequency of oscillation

Fig. 11.27   Making a transverse wave on a long spring

How do the standing waves shown in the photographs arise at all, and why only at particular frequencies? You may remember, in earlier work (for example, *Physics in Action* Chapter 7), seeing a wave travelling along a stretched spring, as in Fig. 11.27. When the wave reaches the end of the spring, it rebounds, changing sides (changing **phase**) as it does so. Suppose the wave arrives back at the starting end and reflects (changing phase as it does so) just as a new wave sets out. The two waves will *reinforce* each other to give a bigger wave. This process repeats itself and a large amplitude builds up.

You can see that this will only happen if the next wave sets out at exactly the right instant, that is, if the frequency is just right.

Fig. 11.29 shows the sequence of events for the standing wave illustrated in Fig. 11.23. Notice that the first wave sets out, rebounds, and meets the second wave halfway down the string. They pass through each other but, since they are out of phase, they cancel at this point, and so the resulting wave has no amplitude in the middle of the string. The first wave reaches the start again just as the third wave is setting out; they meet the second wave on its way back in the middle of the string again, and so on.

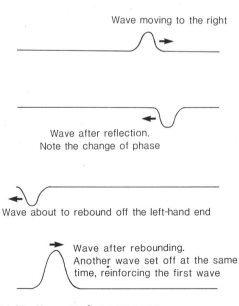

Fig. 11.28   How a standing wave occurs

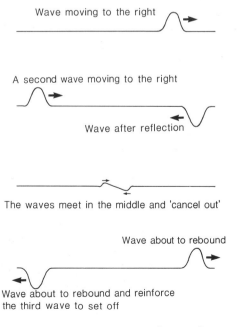

Fig. 11.29   Forming a standing wave with two nodes

11.7 How does sound reach your ear?

When a musical instrument (or anything else) vibrates, it knocks into the air molecules next to the instrument and sets them in motion. These in turn knock into the next air molecules, and so on. A sound **wave** travels out from the vibrating instrument.

This process can be demonstrated using a Slinky spring, as in Fig. 11.30. You may well have seen this before. If one end is vibrated as shown in the next diagram, each coil pushes then pulls its neighbour, and a compression wave travels down the spring. Each coil is oscillating along the direction of the wave in a similar way to the oscillators you looked at earlier in this chapter. This is an example of a **longitudinal wave**.

Sound waves behave in a similar way. The atoms or molecules carrying the sound wave vibrate backwards and forwards so that a series of **compressions** and **rarefactions** travel through the substance (Fig. 11.32).

Notice that this explanation means that sound waves need something to travel through. They will not go through a vacuum. Design an experiment to test this.

It is not easy to draw a diagram like Fig. 11.32, so we usually represent a sound wave in the way shown in the next diagram. The y-axis represents the **displacement** of the particle from its mean position, but you must remember that the particles are really travelling backwards and forwards in the same direction in which the wave is moving. The x-axis represents the distance along the path of the wave.

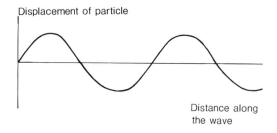

Fig. 11.33 A wave diagram

Fig. 11.30 Making a longitudinal wave on a Slinky spring

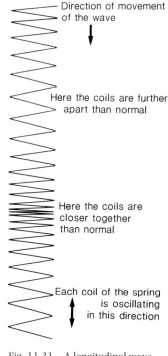

Direction of movement of the wave

Here the coils are further apart than normal

Here the coils are closer together than normal

Each coil of the spring is oscillating in this direction

Fig. 11.31 A longitudinal wave

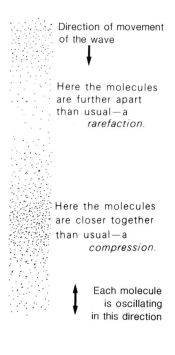

Direction of movement of the wave

Here the molecules are further apart than usual—a *rarefaction*.

Here the molecules are closer together than usual—a *compression*.

Each molecule is oscillating in this direction

Fig. 11.32 How a sound wave travels

The **wavelength** of a sound wave is the distance between successive compressions. The **frequency** of the wave is the number of compressions that pass a given point every second. This, of course, will be the same as the frequency of the oscillator emitting the waves.

If, in 1 second, f waves pass by a given point with a distance λ between each wave, the total length that a wave will have travelled in that second is $f\lambda$. But the distance anything travels in 1 second is the velocity, so the velocity of a wave is the frequency multiplied by the wavelength:

$$v = f\lambda$$

11.8 Measuring the speed of sound

As usual, to measure the speed of anything you need to measure the *time* taken for the sound to travel a measured *distance*. How do you calculate the speed once you have measurements of distance and time?

You can obtain a rough idea of the speed of sound by watching a cricket match. You see the batsman hit the ball, and a short time later you hear the thud of the ball hitting the bat. Light travels very fast indeed, so the delay between seeing and hearing the bat hit the ball is the time taken for the sound to travel from the batsman to you.

If you are watching a cricket match, about how far away from the batsman would you expect to be? Roughly how long is the time delay between seeing and hearing the ball hit the bat? What is a very rough value for the speed of sound?

Suppose there is not a handy cricket match for you to watch. How would you carry out an experiment over a fairly large distance to attempt to measure the speed of sound?

You will agree that, as usual, the difficult part is accurately measuring the time. Measuring a large distance should not be a particular problem. If you have to carry out an experiment in a lab, then the time that you will be trying to measure will be very short indeed, and you will need a suitable clock. The experiment outlined below uses an oscilloscope as a clock; an oscilloscope can easily measure times as small as 1 microsecond (μs), i.e. 10^{-6} second.

The apparatus is illustrated in Figs 11.34 and 11.35. The oscilloscope timebase is switched on so that the spot moves across the screen at regular intervals. Every time the spot starts to move across the oscilloscope screen, a pulse of electricity emerges from the 'X OUT' socket. This pulse is amplified and fed into a loudspeaker, where it is converted to a 'blip' of sound. This sound travels across to the microphone, which converts the sound back into an electric pulse which causes the oscilloscope trace to oscillate as shown. The time delay 't' (Fig. 11.35) represents the time for all this to happen.

The time taken for the sound to travel from the loudspeaker to the microphone is only *part* of the time delay t. What is happening for the rest of the time?

If the microphone is now moved further from the loudspeaker, the sound will have to travel *further* in the air, so there will be an *extra* time delay and t will be longer. If you divide this *extra* distance by the *extra* time, you will obtain the speed of sound in air.

Fig. 11.34   This apparatus can be used to measure the velocity of sound in air

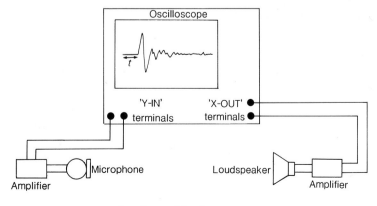

Fig. 11.35   Measuring the velocity of sound

To measure the time, you need to know how fast the spot moves across the oscilloscope screen. The 'timebase speed' control on the oscilloscope tells you this. As an example, look at Fig. 11.36, in which you can see an oscilloscope trace for this experiment, together with all the oscilloscope controls. Look at the control at the bottom right marked 'timebase speed'. This tells you how many milliseconds or microseconds it takes for the spot to travel 1 cm across the screen. In this case the spot takes 1 ms to move 1 cm across the screen.

Look at the oscilloscope screen. You can see the blip representing the arrival of the sound at the microphone. The next photograph shows the screen when the microphone had been moved further from the loudspeaker. The blip is now further from the start of the trace. How much further (in cm)? How much extra time does this represent? What else do you need to know before you can calculate the speed of sound?

You will have to carry out the experiment for yourself to obtain this extra information!

11.9   Wind instruments

Fig. 11.38
There is a standing wave being made in this clarinet

Fig. 11.36
A close view of the oscilloscope screen from Fig. 11.34

Fig. 11.37
The same screen with the microphone further from the speaker

Just as with a stringed instrument, the clarinet that Kate is blowing in Fig. 11.38 has a standing wave set up in it; a standing wave of oscillating air molecules in this case.

The simplest standing wave that can be set up in a closed tube is shown symbolically in Fig. 11.39. (Remember that the air molecules are really vibrating along the length of the tube.) There is a **node** at the closed end—the air molecules cannot vibrate there (Why not?) and there is an **antinode** (a point of maximum vibration) at the open end. By looking at Fig. 11.40, you can see that the wavelength is four times the length of the tube.

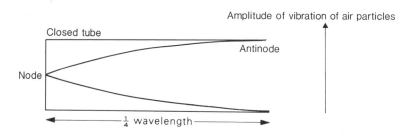

Amplitude of vibration of air particles

Closed tube

Antinode

Node

$\frac{1}{4}$ wavelength

Fig. 11.39
A standing wave in a closed tube, such as the clarinet

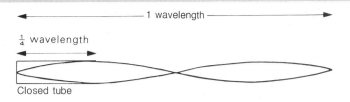

1 wavelength

¼ wavelength

Closed tube

Fig. 11.40   The fundamental wavelength is four times the length of the tube

Just as with the stringed instrument, overtones are possible, as shown in Fig. 11.41. The overtones have frequencies 3f, 5f etc. (where 'f' is the fundamental frequency). A string has a different set of overtones (What were they?) so you would expect a note on a wind instrument to sound different from the same note on a stringed instrument. Fig. 11.43 shows oscilloscope traces of a clarinet and, for comparison, a violin. (These were obtained in a similar way to those for Fig. 11.24.)

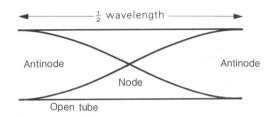

½ wavelength

Antinode   Antinode

Node

Open tube

Fig. 11.42   The standing wave for the fundamental of an open pipe

Overtone with frequency 3f

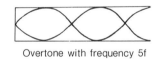

Overtone with frequency 5f

Fig. 11.41   How overtones occur in a closed pipe

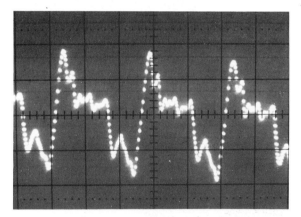

Kate's clarinet is about 0.6 m long. The wavelength of the fundamental note is therefore 2.4 m. Since the speed of sound is 330 m/s, the fundamental frequency is given by

$$f = \frac{v}{\lambda}$$

$$= \frac{330\,\text{m/s}}{2.4\,\text{m}} \quad = 137.5\,\text{Hz}$$

The lowest note from a clarinet is actually the D below middle C, with a frequency of 147 Hz. (A clarinet is a little more complicated than the simple tube in Fig. 11.39!)

Kate can make different notes by opening different holes in the instrument, thus changing the effective length of the air column.

How does an organ produce different notes? What about a trumpet or a trombone?

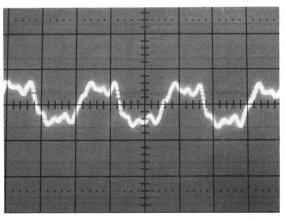

Fig. 11.43   The waveforms from a violin and a clarinet, shown on an oscilloscope screen. Both instruments were playing the same note

The clarinet, together with all other wind instruments, is another example of a resonating system. By blowing into the instrument, Kate vibrates the flexible **reed** at the mouthpiece. This vibrates at a mixture of many frequencies, one of which will be the frequency of the note being fingered, i.e. the resonant frequency of the system, so this frequency will be amplified by the instrument.

Try to find out how an organ, trombone, and flute set their air columns vibrating.

Why do you think stringed instruments usually need a sound box (or sounding board, as in a piano), whereas woodwind and brass instruments do not?

11.10   How do you hear sounds?

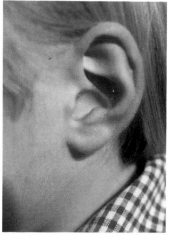
Fig. 11.46

Your ear is yet another system which relies on resonance. Look at the diagram of the ear. Sound waves cause the **eardrum** to vibrate in step with the

Fig. 11.44   The reed of a clarinet

Fig. 11.45
Organ pipes

waves. In the middle ear an arrangement of three bones (called the **hammer**, **anvil** and **stirrup**) transmit these vibrations to the oval window on the **cochlea**. This spiral passage is full of liquid, and there is a flexible membrane along the entire length, in which are hairs connected to nerve endings which lead to the brain. It is thought that different parts of the membrane, or different hairs, have different resonant frequencies, so when a vibration is received the brain can distinguish the different frequencies by the part of the membrane which resonates.

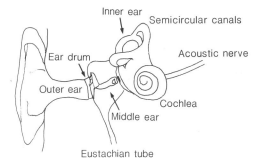

Fig. 11.47   Inside an ear

SUMMARY

Now that you have finished studying this chapter on oscillations and sound, there are a number of things you should know and be able to do.

1 You should:
   a) know what is meant by the frequency of an oscillation;
   b) know the units of frequency;
   c) know the factors which affect the period of some simple oscillators;
   d) know the definition of simple harmonic motion;
   e) understand what causes damping;
   f) know the shapes of the displacement-time, velocity-time and acceleration-time graphs for something moving with simple harmonic motion;
   g) understand what is meant by resonance, and how it occurs;
   h) understand how the wind can cause oscillations as a result of vortex shedding;
   l) understand what a standing wave is, and how it occurs;
   j) understand how overtones occur in musical instruments;
   k) know how a sound wave travels;
   l) be able to describe one experiment to determine the speed of sound.

2 You should know what each of the following is, or what it does:

pendulum	oscillation
compression wave	period (of oscillation)
amplitude	mean position
quality (of note)	displacement
sine wave	eddy
vortex	forced oscillation
fundamental frequency	timbre
	pulse
rarefaction	antinode
node	

3 These are some of the other words that have been used in this chapter. You should know what each word means:

condition	formal
twisting	destructive
natural	streamlined
obstruction	absorb
reinforce	

FURTHER QUESTIONS

Assume that the velocity of sound in air is 330 m/s.

1 Jenny whistles into a microphone connected to an oscilloscope. The trace on the oscilloscope is as shown in Fig. 11.48(a). Tom then whistles into the same microphone from the same distance. The trace on the oscilloscope is then as in Fig. 11.48(b). The controls on the oscilloscope have not been adjusted in any way between Tom and Jenny whistling into the microphone.
a) Who whistles the louder?
b) Who whistles with the greater frequency?
c) Which note has the greater wavelength?

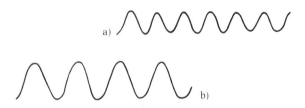

Fig. 11.48

2 Suggest a reason for the following:
a) as a nail is hammered into a wall, the pitch of the note made by the hammer striking the nail rises;
b) the humming sound made by a gnat when flying is of a higher frequency than the sound made by a bee.

3 The timekeeper in a 100 m race stands at the finish, and starts a stopwatch on hearing the bang from the starting pistol.
a) Why will the timekeeper measure the incorrect time for the race?
b) Calculate the error that is made in timing this race.

c) When should the timekeeper have started the stopwatch?

4 What is it about a swinging pendulum that makes it useful for regulating a clock?

5 a) If you displace a trolley tethered between springs (as in Fig. 11.5) and release the trolley, it oscillates backwards and forwards. Why does it do this? Why does it not just come back to the rest position and stay there?
b) If the mass of the trolley is increased, the time taken for one oscillation also increases. Explain why.
c) If two springs are used at each end of the trolley instead of one, the time for an oscillation decreases. Explain why.

6 Once when Vicki was standing near a cliff, she noticed she could hear an echo of her voice a short time after she shouted. She thought she could use this echo to measure the speed of sound. She stood exactly 200 m from the cliff and clapped her hands once. She thought that it was about 1 second before she heard the echo.
a) What value does this give for the velocity of sound?
b) Why do you think Vicki had difficulty in measuring the time accurately?
Vicki then tried clapping continuously, giving a clap as soon as the echo from the previous clap returned to her. She found she took 12 seconds to make 10 claps.
c) How long did it take the sound from one clap to travel to the cliff and return to Vicki?
d) How fast was the sound travelling?
e) Why do you think Vicki's second method gives a more reliable result than her first measurement?

12

WAVES AND LIGHT

Fig. 12.1   The spectrum of sunlight

12.1   Introduction

You will recognize the photograph above as consisting of all the colours of the spectrum of the light from the Sun—the colours of a rainbow. White light is a mixture of all these colours. How were these colours separated in order to make this photograph? It certainly is not a photograph of a rainbow.

You should already know that light is a wave motion. How do scientists *know* that light is a wave motion? The wavelength of light is very small and each colour has its own wavelength. How can these small wavelengths be measured? What values would you expect to obtain?

This chapter is mainly about light waves. In it you will find the answers to these questions. You will also learn that there are other important waves similar to light waves.

12.2   You should know . . .

This chapter assumes you have already investigated some of the ways in which waves behave, mainly using a ripple tank like the one illustrated in Fig. 12.2. If you have not, you should read *Physics in Action* Chapter 7 before reading this chapter.

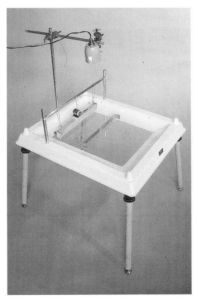

Fig. 12.2   A ripple tank, for studying
waves on water

To understand this chapter you should know:
1 how to set up and use a ripple tank, and how to use a stroboscope to 'freeze' the movement of waves on the tank;
2 the meaning of frequency, wavelength and velocity of a wave;

123

3 that waves reflect from a barrier so that the incident angle is equal to the reflected angle (Fig. 12.3);

4 the meaning of refraction, diffraction and interference.

Some of these points were covered in the last chapter. Here is a brief reminder of some of the others.

Waves change their speed as they pass from one substance to another, or, in the case of water, as the depth of water changes. This can cause a wave to change direction, as shown in Fig. 12.4. This process is known as **refraction**.

Waves spread out as they pass through a small gap, as shown in Fig. 12.5. This is called **diffraction**. It is most noticeable if the gap is no more than a few wavelengths wide.

Waves can add together to give a larger amplitude, or can cancel, to give no wave (Fig. 12.6). You will have already met this idea if you have looked at standing waves in Section 11.6. This is called **interference** of waves.

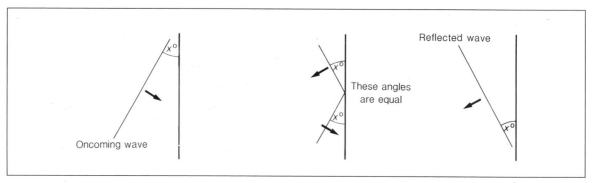

Fig. 12.3   The reflection of waves

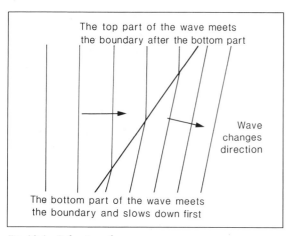

Fig. 12.4   Refraction of waves

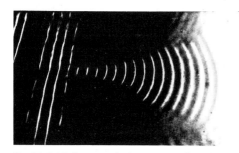

Fig. 12.5   Diffraction of waves on a ripple tank

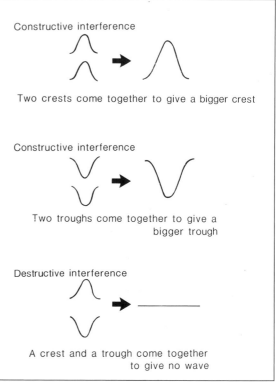

Fig. 12.6   Adding waves together

12.3 Interference of water waves

You will probably have done this experiment before, but it is very important that you understand it, so it is repeated here.

Assemble a ripple tank in the usual way. Hang two circular wave makers in the water at one end of the tank, as in Fig. 12.7. When you switch on the motor, two identical sets of waves will travel down your ripple tank. Obviously, they will overlap as they spread down the tank. What happens when they overlap? Look at Fig. 12.8. This photograph shows two sets of overlapping waves, and you can see that there are some paths along the tank where there *are* waves, and other paths where there are no waves.

Try to obtain this pattern with your tank. Does changing the wavelength change the pattern? Suppose you use two wave makers that are further apart. Does this make any difference to the pattern?

How can we explain this pattern? How can two sets of waves join together to give places where there are no waves at all?

Remember that water waves consist of crests and troughs following each other. Suppose, at some place in your tank, a crest from one vibrator arrived at the same time as a crest from the other vibrator. These two crests would pile one on top of each other to form a bigger crest as in Fig. 12.6. A small time later a trough will arrive from one rippler at the same time as a trough from the other rippler. These will add together to give a deeper trough.

Waves from one rippler are adding to, and re-inforcing, the waves from the other rippler. This is **constructive interference** of waves.

Fig. 12.7   These two vibrators are used to make two sets of circular waves

Fig. 12.8   An interference pattern formed by two sets of circular waves on a ripple tank

In another part of the tank, a crest from one rippler might arrive at the same time as a trough from the other rippler. The crest will 'fill in' the trough, the waves will cancel and give flat water. This is **destructive interference**.

Fig. 12.9 explains how the pattern over a large area of your ripple tank occurs. This pattern is called an **interference pattern**. It is important that you understand how this interference pattern occurs, before you try to understand the rest of this chapter.

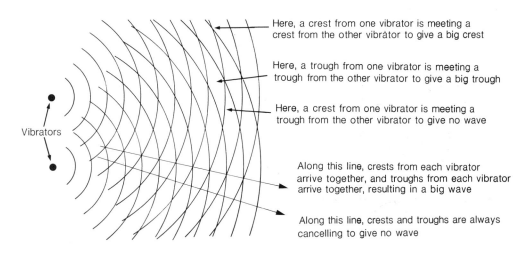

Vibrators

Here, a crest from one vibrator is meeting a crest from the other vibrator to give a big crest

Here, a trough from one vibrator is meeting a trough from the other vibrator to give a big trough

Here, a crest from one vibrator is meeting a trough from the other vibrator to give no wave

Along this line, crests from each vibrator arrive together, and troughs from each vibrator arrive together, resulting in a big wave

Along this line, crests and troughs are always cancelling to give no wave

Fig. 12.9   How an interference pattern is formed

12.4 Are microwaves really waves?

Fig. 12.10 A microwave generator

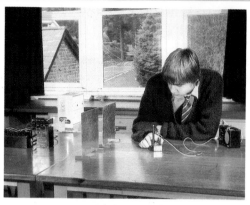

Fig. 12.11 Investigating the diffraction of microwaves

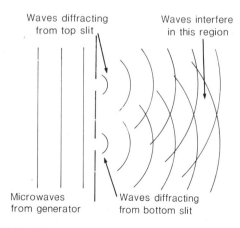

Fig. 12.12 Investigating the interference of microwaves

You probably know that the box in Fig. 12.10 gives out waves. The waves are invisible; they are like radio waves with a very short wavelength and are often called **microwaves**. (These are not the same microwaves as are used in a microwave oven, although they do belong to the same 'family' of waves.)

How do we *know* waves are being given out by the box in Fig. 12.10? Only waves diffract and interfere, or, to put this the other way round, if we can obtain diffraction or an interference pattern, this must be caused by waves. Particles, like footballs, cannot spread out through gaps, nor can they add together to make bigger footballs, or cancel out.

You can look for *diffraction* of microwaves using the apparatus illustrated in Fig. 12.11. The gap through which the waves pass should be about 3 cm wide. The rod that Richard is holding detects the microwaves and converts them to electrical energy. This is fed into an amplifier and loudspeaker which changes the energy into sound so that you can detect the position of the microwaves using your ears. If you move the detector across the bench 0.5 m from the gap, what do you hear? Are microwaves spreading out behind the gap? Do you think diffraction is taking place?

To try to obtain an interference pattern with microwaves you might think you need *two* microwave generators. But this is not necessary; the arrangement Sheila is using in Fig. 12.12 works very well. The waves from the microwave generator pass through the two gaps and diffract (or spread out) as in Fig. 12.13. In effect they are behaving *as if* they are coming from two sources, one in the middle of each gap.

Fig. 12.13 Microwaves passing through a gap

Set up this apparatus and push the detector across the bench 0.5 m from the gaps in the same way as you did when looking for diffraction. Do you think there is an interference pattern there?

12.5 What is the wavelength of microwaves?

The interference experiment that Sheila was illustrated doing in the last section can be used to *measure* the wavelength of microwaves. Look at Fig. 12.14. There is a *maximum* amplitude at the point marked 'M'. Since the waves from one slit have exactly the same distance to travel to M as the waves from the other slit, the waves from each slit arrive at M *in phase*, that is, a crest from one slit arrives at the same time as a crest from the other slit, followed by a trough from each slit arriving simultaneously.

Suppose the next point where there is a maximum amplitude is at P. The waves from each slit must also be arriving in phase here. But it is obviously further from the bottom slit to P than from the top slit to P. Since the waves *leave* the slits in phase (Why?) and arrive at P in phase, it must be possible to fit exactly one whole extra wavelength into the path from the bottom slit to P, compared to the path from the top slit to P. The diagram shows eight waves in the top path and nine waves in the bottom path. We say the **path difference** is one wavelength.

If the next maximum is at Q (i.e. the second maximum counting from the centre), how many extra waves fit into the 'bottom' path to Q compared to the 'top' path to Q? What is the path difference this time?

To measure the wavelength of the microwaves, set up the apparatus illustrated in Fig. 12.15 on a large sheet of paper. Draw a line across this paper about 0.5 m from the gaps. Move the detector along this line until you find the central maximum. Mark this position. Then find the position of the first maximum to one side of the centre. Mark the position of this first maximum. Measure the distance from each slit to the position of this maximum. The difference between the two distances (i.e. the path difference) is equal to one wavelength.

Try to find the position of the second maximum, and measure the path difference in the same way. How many wavelengths is *this* path difference equal to?

Fig. 12.15 Sheila measuring the wavelength of microwaves

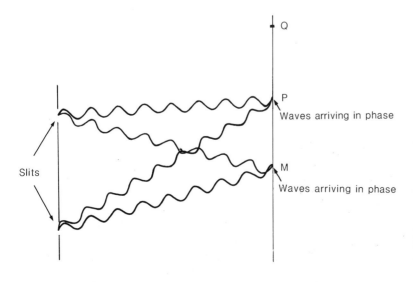

Fig. 12.14
How waves form maxima in an interference pattern

127

12.6   Light waves

As you already know, light is a wave motion. However, you do not normally notice this, and you have to look very carefully for evidence that light consists of waves. Have you any idea why this might be so?

A 'double slits' experiment similar to that described for microwaves in the previous section shows up the wave nature of light, but the *scale* of the apparatus has to be changed considerably. The slits should be about $\frac{1}{4}$ mm wide and $\frac{1}{2}$ mm apart (measured from centre to centre of the slits) as shown in Fig. 12.16. If you have to make the slits for yourself, a good way to do it is to paint a microscope slide with 'Aquadag'; allow the paint to dry and then use a very sharp craft knife (or a single-sided razor blade) and a steel rule to cut two slits about $\frac{1}{2}$ mm apart. Take care to make the slits parallel to each other.

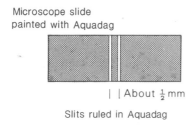

Microscope slide
painted with Aquadag

| | About $\frac{1}{2}$ mm

Slits ruled in Aquadag

Fig. 12.16   A double slit

Figs 12.17 and 12.18 should help you to set up this experiment for yourself. To be successful the experiment must be done in a darkened room. It will probably be necessary to use makeshift screens to prevent scattered light reaching the screen. If there are several experiments in the same room,

it will need careful organization to prevent light from one experiment scattering onto the screen of another.

The lens is to concentrate the light and give a sharp pattern on the screen. Place the lamp, with its filament vertical, at one end of a bench, and a screen (a ground glass one is best) about 3 m away. Put a weak converging lens (about +2 dioptres) midway between the two, and slide it towards the *lamp* until you obtain a sharp, enlarged image of the lamp filament on the screen. Two people are needed to do this, one to move the lens and the other to watch the screen.

If you slide the lens towards the *screen* instead of the lamp, you will probably obtain a *smaller* image of the filament at some point. This is *not* what you want.

When you have a sharp image, put the double slits in front of the lens as shown. Look for an interference pattern on the screen. What does the pattern look like?

Fig. 12.18   Young's double slit experiment

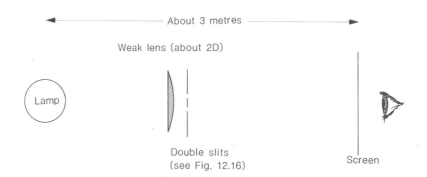

About 3 metres

Weak lens (about 2D)

Lamp

Double slits
(see Fig. 12.16)

Screen

Fig. 12.17   Young's double slit experiment

You should appreciate that this experiment would never work if light waves did not diffract as they passed through each slit. Without diffraction the two beams of light would carry on in two straight, narrow lines and never meet [Fig. 12.19(a)]. Even with slits as narrow as $\frac{1}{4}$mm the light is not diffracted very much, because the wavelength of light is much less than $\frac{1}{4}$mm. This is why the slits must be narrow (so there is as much diffraction as possible) and close together (to give the light waves a chance of overlapping).

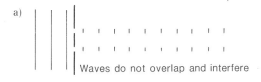

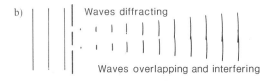

Fig. 12.19   Waves must diffract in order for Young's double slit experiment to work

If you cannot obtain any sort of interference pattern on the screen, then it is probable that your slits are either too wide, or they are too far apart.

This experiment is called the 'Young's double slits' experiment, after the physicist who first performed it in 1801.

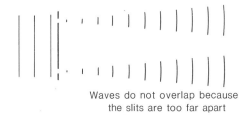

Waves do not overlap because the slits are too far apart

Fig. 12.20   These slits are too far apart; the waves do not overlap

At the time when Young performed this experiment there was some argument among scientists as to whether light consisted of a wave motion, or consisted of a stream of tiny particles, or 'corpuscles'. (Newton favoured the 'corpuscular theory', as it was called.) Young's experiment was an important piece of evidence in favour of the wave theory.

You will learn later that the wave theory of light is not quite sufficient to describe all that light does, and that Newton was not totally wrong with his corpuscular theory.

An alternative way of performing the double slits experiment is to *measure* the intensity of the light in the various parts of the interference pattern. This can be done using a **photodiode** in a suitable electric circuit so that the output voltage is proportional to the intensity of light falling on the photodiode. In Fig. 12.21 a VELA was being used to record the output from the photodiode circuit at regular intervals as the photodiode was pushed across the interference pattern at a steady speed. The readings which were stored in the VELA were then output to an oscilloscope. The oscilloscope trace in Fig. 12.22 shows how the intensity of the light varies across the interference pattern.

Why do you think it was important to move the photodiode across the interference pattern at a constant speed?

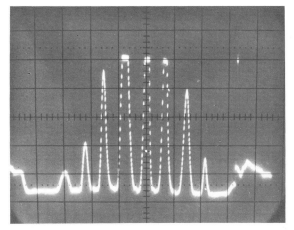

Fig. 12.21   Recording an interference pattern using a photodiode. A laser was used to produce the red light

Fig. 12.22   An oscilloscope trace showing the results of the last experiment

12.7   The wavelength of light

It should be obvious to you that you cannot measure the wavelength of light in the same way as you measured the wavelength of microwaves in Section 12.5, that is, by measuring with a ruler the distance from each slit to a maximum, and subtracting the two distances. Why is this not possible?

However, it is possible to use Young's double slits to measure the wavelength of light, like this. Look at Fig. 12.23. S_1 and S_2 represent the two slits, separated by a distance s. On the screen, there is a maximum in the *middle*, at C, since the waves from S_1 and S_2 both have to travel the same distance and so arrive at C in phase.

Suppose the first maximum to one side of C is at B, a distance d from the centre. The wave from S_2 clearly travels further than the wave from S_1. Since the waves arrive in phase at B (since there is a maximum intensity here, which is caused by constructive interference) the wave from S_2 must travel an extra distance exactly equal to one wavelength.

Why *one* wavelength? Why not two, three or four wavelengths?

The extra length of the path S_2 is the distance S_2Z. This is the distance we were able to measure with a ruler when dealing with microwaves. In this case we must find a way of *calculating* it.

Look at the triangle ABC. This is a right-angled triangle. Triangle S_1S_2Z is almost the same shape as triangle ABC; angle a = angle b (Why?) and angle z is almost equal to a right angle.

Since these triangles are the same shape, the *ratio* of any two sides in one triangle is the same as the ratio of the corresponding two sides in the other triangle. Therefore:

$$\frac{BC}{AC} = \frac{S_2Z}{S_1Z}$$

If you prefer, you can say that since

$$\text{angle } a = \text{angle } b$$
$$\tan a = \tan b$$
$$\frac{BC}{AC} = \frac{S_2Z}{S_1Z}$$

BC is the distance from one maximum on the screen to the next (the **fringe separation**) and is equal to the distance d. AC is the distance from the slits to the screen and is equal to L. We have agreed that S_2Z is one wavelength (λ), and S_1Z is almost equal to the distance between the slits, which is s. Therefore we can rewrite the equation as

$$\frac{d}{L} = \frac{\lambda}{s}$$

or

$$\lambda = \frac{ds}{L}$$

or

$$\text{wavelength} = \frac{\text{fringe separation} \times \text{slit width}}{\text{distance from slits to screen}}$$

We have had to make two very small approximations in deriving this equation. What are they?

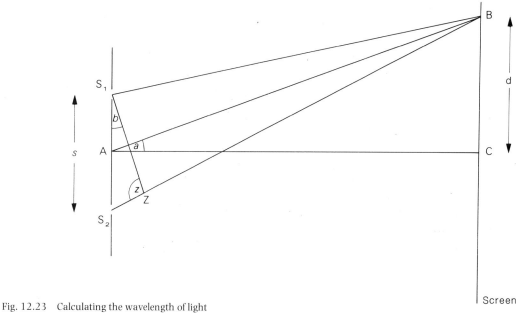

Fig. 12.23   Calculating the wavelength of light

These approximations may not seem particularly small to you, but remember that Fig. 12.23 is not drawn to scale. The distance L is greater than 1 m, while the distance d is probably less than 1 mm.

Try to measure the average wavelength of light using the apparatus you assembled for the last experiment to observe the interference of light. You must measure d, s, L so that you can substitute these values into the equation above (make sure you get the units right!)

Measuring L should cause you no problem. To measure d, measure 5 or 10 fringe separations and then divide this distance by 5 or 10.

Measuring s will probably cause you the greatest problem. You may be lucky and have a ready ruled pair of slits of known separation, or be able to use a measuring or travelling microscope so that you can make this measurement. Alternatively, put your slits up against a transparent ruler, preferably one marked at $\frac{1}{2}$ mm intervals, on an overhead projector; you should be able to make an intelligent estimate of the slit separation from the enlarged image of the slits and ruler.

12.8 The diffraction grating

Compare Figs 12.24 and 12.25. Both show interference fringes formed by passing light through double slits as explained in the last section. Green light was used to make the fringes in Fig. 12.24. White light was used in Fig. 12.25. Although green light gives green fringes, white light does not give pure white fringes. Notice that they are *coloured*, especially at the edges, with the blue edge nearest the centre of the pattern and the red end furthest from the centre.

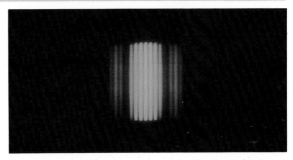

Fig. 12.24 An interference pattern formed with green light

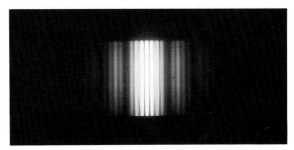

Fig. 12.25 An interference pattern formed with white light. In what way does it differ from the previous pattern?

These coloured edges are easy to explain if you remember that different *colours* have different *wavelengths*. If we rewrite the double slits equation from the last section as

fringe separation

$$= \frac{\text{wavelength} \times \text{slit-screen distance}}{\text{slit separation}}$$

you can see that the *fringe separation* depends on the *wavelength*, the larger the wavelength, the larger the fringe separation. So different wavelengths will form maxima at different positions on the screen.

Fig. 12.26 shows an alternative way of explaining the separation of the colours.

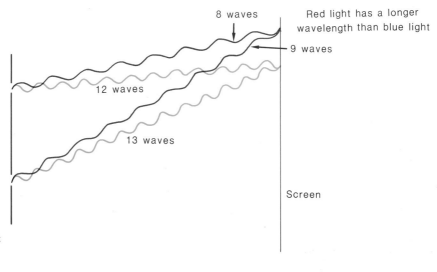

8 waves

9 waves

12 waves

13 waves

Red light has a longer wavelength than blue light

Screen

Fig. 12.26 Why different colours form maxima at different places

The colours would be better separated and so easier to see if the slits were closer together. But slits which are close together must be very narrow slits, or the slits will overlap and become one wider slit. (Remember that the slit separation is measured from centre to centre.) Narrow slits do not let much light through, so the interference pattern would be very dim indeed.

We can obtain a brighter pattern if we use more than two slits. The interference pattern in Fig. 12.27 was formed by passing light through 100 parallel slits. You can see the pattern is bright, and also much *sharper* than that formed with just two slits. This idea can be extended further; the slits can be made very narrow so they can be very close together, separating the maxima much further. To make the pattern in the next photograph, light waves passed through 2000 parallel slits, each slit being 10^{-5} m from its neighbour. The colours are now very well separated. These slits are made by ruling furrows in a piece of glass. Plastic mouldings, or 'replicas' are then made of this glass 'original'. The furrows scatter light and act as the opaque parts between the clear (unmarked) glass 'slits'. This device is known as a **diffraction grating**.

The spectrum at the start of this chapter was made by shining sunlight through a very fine grating with a slit separation of 10^{-6} m. Diffraction gratings are very useful for examining the spectrum (that is, the different wavelengths) given out by hot

Fig. 12.27   The spectrum of white light seen through a diffraction grating of 100 slits

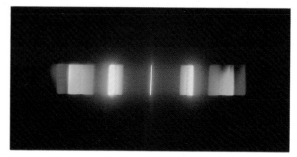

Fig. 12.28   The same spectrum through 2000 slits

gases. Each element gives out its own particular set of wavelengths. Fig. 12.29 shows **emission spectra** of some common elements. Elements present in stars have been identified by **spectral analysis** of the light from the stars. The spectrum from an element gives important information about the *structure* of the atom.

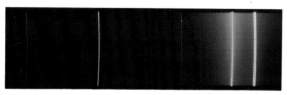

(a) Hydrogen

(b) Helium

(c) Cadmium

(d) Mercury

Fig. 12.29   The emission spectra of some common elements

12.9 The electromagnetic spectrum

As you know, both the 3 cm 'microwaves' and light waves belong to a large family of waves called **electromagnetic waves**. The members of this family, together with the corresponding wavelengths and other useful information, are illustrated in Fig. 12.30. Notice the very wide range of wavelengths in this spectrum, from radio waves with wavelengths greater than 1 km to gamma rays with wavelengths less than the diameter of an atom.

Despite the wide range of wavelengths, all the waves travel at the same speed; in free space this speed is 3×10^8 m/s.

Electromagnetic waves are produced when charged particles change their energy. This change in energy could, for example, be the result of a change of energy of an electron within an atom, or an electron oscillating in a radio transmitting aerial. The greater the energy change, the higher the frequency of the electromagnetic radiation.

Your eyes can only see a very small part of this spectrum. The diagram shows how some of the other wavelengths are detected.

You will learn more about gamma rays in Chapter 20.

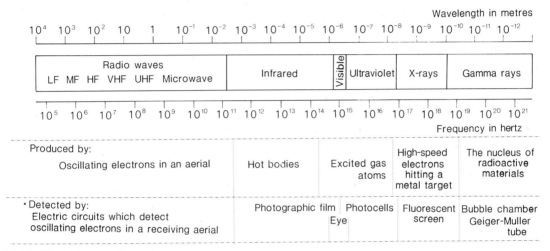

Fig. 12.30 The electromagnetic spectrum

SUMMARY

Now that you have finished studying this chapter on waves and light, there are a number of things you should know and be able to do.

1 You should:
   a) be able to use a ripple tank to study waves on water;
   b) understand what happens to waves when they
      i) reflect
      ii) refract
      iii) diffract
      iv) interfere;
   c) be able to explain how an interference pattern from a double slits experiment arises;
   d) know how to set up a double slits experiment for both microwaves and light;
   e) know how to use a double slits experiment to measure the wavelengths of both microwaves and light;
   f) understand and be able to explain how a diffraction grating works;
   g) know the various parts of the electromagnetic spectrum.

2 You should know what each of the following is, or what it does:

spectrum	interference pattern
constructive	destructive
interference	interference
crest	path difference
trough	rippler
microwave	amplitude

3 These are some of the other words that have been used in this chapter. You should know what each word means:

identical	cancel
scale	makeshift
scatter	corpuscle
absorb	corresponding

FURTHER QUESTIONS

1 If light diffracts, why can you not see round corners?

2 Explain why it is necessary for light to diffract in order that Young's double slits interference fringes can be formed.

3 Your school laboratory is probably illuminated by several lamps. Why, when they are all switched on, does the light from the various lamps not form an interference pattern on the floor or walls?

4 White light is used to obtain an interference pattern using Young's double slits. The slits are 0.6 mm apart. The distance between the slits and the screen on which the fringes are formed is 1.8 m, and the distance between successive bright fringes is 1.5 mm.
   a) Each bright fringe has coloured edges, one edge being blue and the other being red; explain why.
   b) Explain whether, for any particular fringe, the red edge will be closer to the centre of the screen, or further from the centre.
   c) Calculate the average wavelength of the white light.

5 Look at Fig. 12.31. These diagrams show waves coming from the left and passing through part of a diffraction grating. The three diagrams show waves travelling in three different directions after passing through the grating.
   a) Explain why the waves travelling in the direction shown in diagram a) will form a 'maximum' (i.e. a bright patch) on a screen to the right of the diagram.
   b) Explain why the waves travelling in the directions shown in diagrams b) and c) will also form maxima on a screen.
   c) Using these diagrams, explain why red light forms maxima in different positions on a screen from blue light.
   d) Draw similar diagrams to those in Fig. 12.31 to explain why a 'coarse' grating (i.e. a grating with slits spaced far apart) gives many more maxima on a screen than a 'fine' grating.

6 A portable radio, fitted with a telescopic aerial (for receiving VHF) is tuned to a radio station transmitting radio waves of frequency 90 MHz. Radio waves travel with a velocity of 3×10^8 m/s in air.
   a) What is the wavelength of these radio waves?
   b) As a person moves across the room in which the radio is situated, it is noticed that the volume of the sound from the radio fluctuates and depends on where the person is standing. Suggest a possible reason for this.

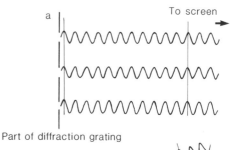

Part of diffraction grating

Fig. 12.31

13

LENSES, MIRRORS AND RAYS OF LIGHT

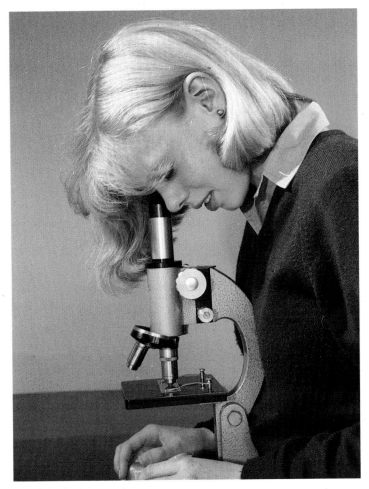

Fig. 13.1   A microscope in use

13.1   Introduction

Why do you think Kate needs to use a microscope in this picture? Obviously, whatever she is looking at is too small to be seen with an unaided eye. The microscope does something to make whatever she is looking at appear larger. To understand how a microscope works, you need to know what the lenses in the microscope do to **rays of light**.

This chapter is about rays of light and how they are put to use in **optical instruments** like the microscope in the photograph. What other optical instruments can you think of?

13.2   You need to know . . .

In order to understand this chapter you need to know:
1 how a pinhole camera works;
2 how a simple lens camera works;
3 how convex and concave lenses form images;
4 the difference between real and virtual images;
5 how a convex lens can be used as a magnifying glass;
6 how to arrange two convex lenses as a simple astronomical telescope;
7 how your eye works, and how long and short

sight are corrected;

8 what is meant by principal focus and focal length of a lens.

If you have never met these ideas before, or forgotten about them, then you should perhaps read *Physics in Action* Chapter 6 before working through this chapter.

13.3 Making a real image with a convex lens

You have probably used lenses before to make an image. You might have made a **real image** on a screen using a convex lens. For example, a slide projector forms a real image of a slide on a screen. A camera uses a lens to form a real image on a film.

Is the real image made with a convex lens the same way up as the 'object', or upside down? Is the image larger or smaller than the object, or can it be either?

What does a convex lens do to light rays to make a real image? You should have investigated this already, perhaps using apparatus similar to that assembled in Fig. 13.2.

Fig. 13.3 shows how a convex lens forms a real image. Two lamps were used, one to represent the top of an object and the other to represent the

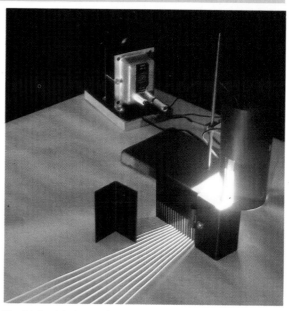

Fig. 13.2   A light rays kit

bottom. The rays from the lamps have been coloured in the photograph so that you can easily follow them. These rays are converged by the lens to form a real image on the right, as shown in Fig. 13.4.

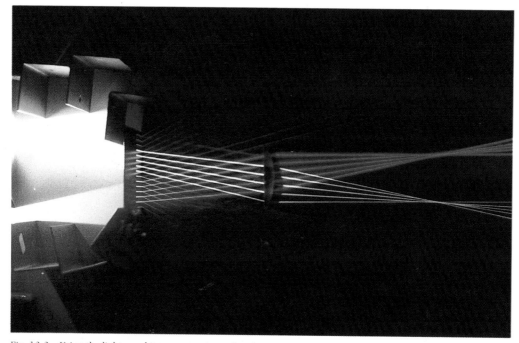

Fig. 13.3   Using the light rays kit to represent rays forming a real image

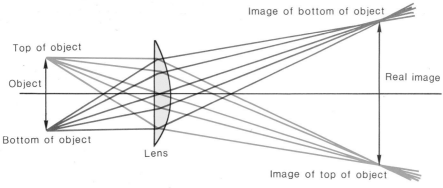

Fig. 13.4 A ray diagram for Fig. 13.3

A diagram like Fig. 13.4 obviously cannot be drawn quickly or easily. Physicists therefore usually draw a simplified diagram. If the focal length of the lens is known, this simplified diagram can be used to calculate the *size* and the *position* of the image if it is drawn accurately.

To draw an accurate diagram, you need to know exactly where some of the rays of light go. Look at Fig. 13.5. This shows two such rays. You can always predict exactly what happens to these rays. One ray leaves the object parallel to the horizontal line through the middle of the lens which is called the **principal axis**, and the other is the ray that goes through the middle of the lens. The diagram shows what happens to them.

13.4 A practical example

To see how to construct an accurate diagram, let us look at a practical example. Fig. 13.6 shows a camera ready to take the photograph of the reed of the clarinet which forms Fig. 11.44. The centre of the lens was 20 cm from the reed. The focal length of the lens was 5 cm. The reed was 1 cm long. How big was the image on the film, and how far was the centre of the lens from the film?

(It does not matter that the lens was actually made up of several 'elements' or individual lenses; we can treat it as one lens for this example.)

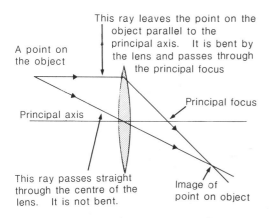

Fig. 13.5 Two important rays passing through a convex lens

Fig. 13.6 A camera ready to take the photo forming Fig. 11.44

Fig. 13.7 shows you the stages in constructing an accurate scale diagram. You may find it easier if you use graph paper. You must decide on suitable horizontal and vertical scales. To draw Fig. 13.7 a horizontal scale of 1 cm on the diagram = 2 cm in reality (i.e. half size) enables the diagram to fit sensibly onto the page. A vertical scale of 1 cm on the diagram = 0.5 cm in reality (i.e. double size) was used. The lens is always drawn at least as large as the object, which usually makes it totally out of scale.

Follow Fig. 13.7 carefully, and make sure you can draw such a diagram for yourself.

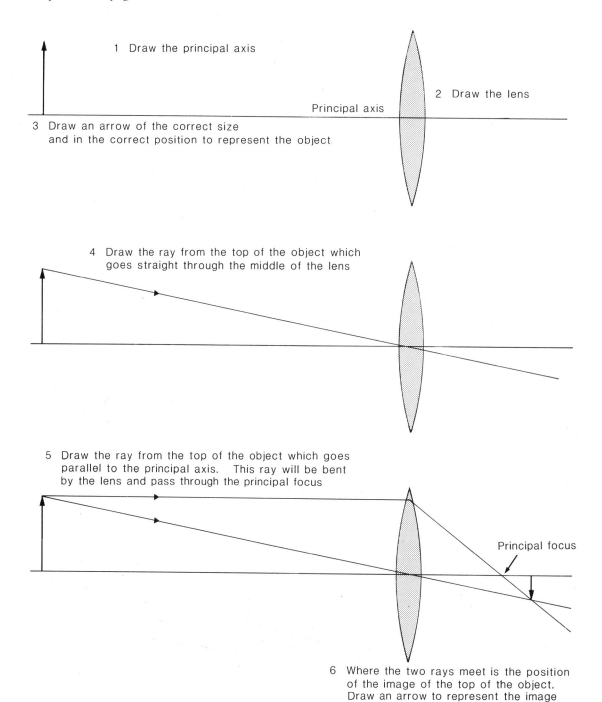

Fig. 13.7   How to draw a scale diagram of a lens forming a real image

13.5 A convex lens as a magnifying glass

Fig. 13.8   Using a magnifying glass

As well as making a real image as described in the last section, a convex lens can be used as a **magnifying glass** as shown in Fig. 13.8. Which way up is this image?

To be used as a magnifying glass, the distance between the lens and the image must be less than the focal length. The image is a **virtual image**.

Figs 13.9 and 13.10 remind you of how this image is formed. The rays of light coming from one point on the object are still spreading out after passing through the lens (but not spreading out as much as before they hit the lens). They *behave* as if they are coming from some point behind the object. If these rays enter your eye, your brain thinks the rays are coming from this point, as shown in Fig. 13.10. This point is a virtual image of the original point on the object.

We can construct an accurate scale diagram in a similar way to the previous one. The lens being used in Fig. 13.8 had a focal length of 5 cm and was being held 4 cm from the map. Where was the virtual image of the word 'Windermere' and how much bigger than the original word was it?

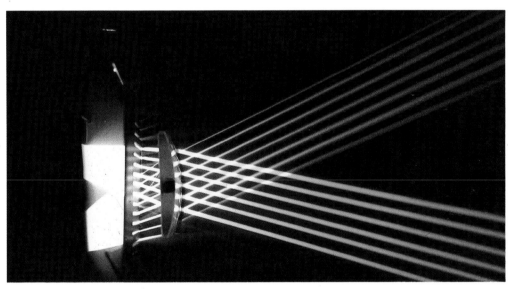

Fig. 13.9   Forming a virtual image with the light rays kit

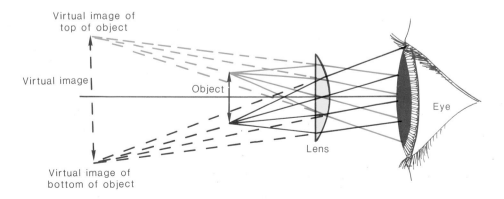

Fig. 13.10   A ray diagram to show how an extended virtual image is formed

Carefully follow Fig. 13.11 to see how a scale diagram can be constructed in order to find the answers to these questions.

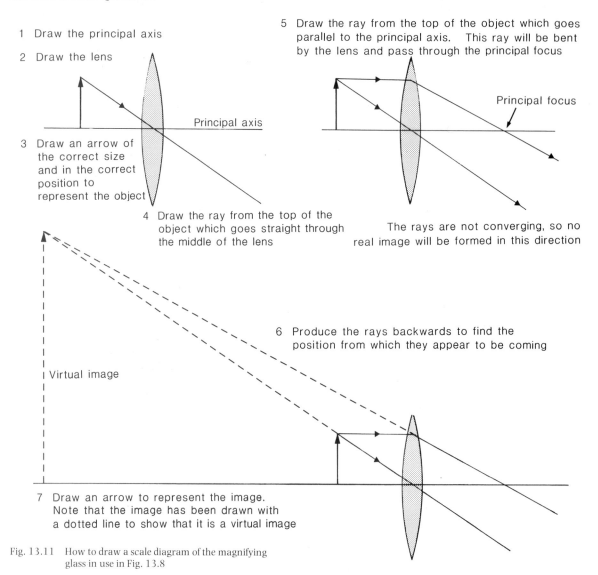

1 Draw the principal axis

2 Draw the lens

Principal axis

3 Draw an arrow of the correct size and in the correct position to represent the object

4 Draw the ray from the top of the object which goes straight through the middle of the lens

5 Draw the ray from the top of the object which goes parallel to the principal axis. This ray will be bent by the lens and pass through the principal focus

Principal focus

The rays are not converging, so no real image will be formed in this direction

6 Produce the rays backwards to find the position from which they appear to be coming

Virtual image

7 Draw an arrow to represent the image. Note that the image has been drawn with a dotted line to show that it is a virtual image

Fig. 13.11   How to draw a scale diagram of the magnifying glass in use in Fig. 13.8

13.6   The focal length of a convex lens

There is a special case with a convex lens when the object is at the principal focus. The rays then emerge from the lens parallel to each other, as shown in Fig. 13.12, so no image is formed. It is sometimes said that the image is 'at infinity', that is, an infinite distance away.

If rays of light which are parallel to each other enter a lens, they will converge to meet at the principal focus, as in Fig. 13.13. (Notice that Fig. 13.13 is the same as Fig. 13.12 in reverse.) This provides the basis of two easy methods of finding the focal length of a convex lens.

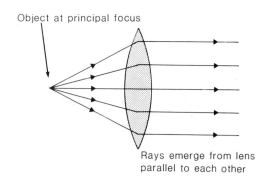

Object at principal focus

Rays emerge from lens parallel to each other

Fig. 13.12   Rays starting from the principal focus are parallel to each other after passing through the lens

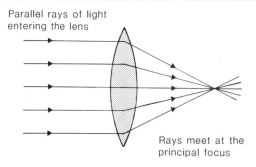

Parallel rays of light
entering the lens

Rays meet at the
principal focus

Fig. 13.13   These parallel rays converge to the principal focus

Fig. 13.15   A second way to measure the focal length of a convex lens

The first method is simply to use the convex lens to form the image of a distant object on a piece of paper, as is being done in Fig. 13.14. A 'distant object' can be a window at the opposite end of your laboratory. The rays from a point on the distant object which pass through the lens are very nearly parallel, so they will converge to form an image in the plane of the principal focus. The distance of the image from the lens will be equal to the focal length. You simply measure this distance from the lens to the image on the piece of paper with a ruler.

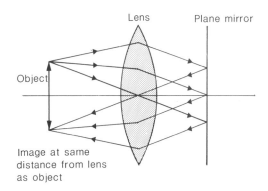

Lens   Plane mirror

Object

Image at same
distance from lens
as object

Fig. 13.16   A ray diagram for Fig. 13.15

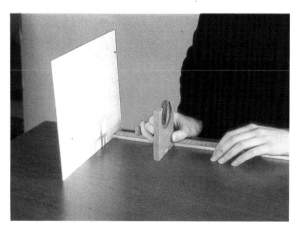

Fig. 13.14   One way to measure the focal length of a convex lens

Do you think this method is very accurate? Which part of the lens do you think you measure to, the middle, or one of the faces?

The second method of finding the focal length of a convex lens is illustrated in Figs 13.15 and 13.16. Crossed wires mounted in a piece of card make a convenient 'object'. The rays of light from the object pass through the lens and hit the plane mirror placed behind the lens. If the distance from the object to the lens is equal to the focal length of

the lens, the rays emerge from the lens parallel to each other, as in Fig. 13.16. They will still be parallel after they have been reflected from the mirror. The rays pass back through the lens and form an image. Since the rays are parallel, the image will be at a distance from the lens equal to the focal length of the lens, that is, at the same distance from the lens as is the object.

To use the apparatus, assemble it as shown in the photograph and slide the lens backwards and forwards until there is a sharp image of the cross-wires on the card. You will probably have to twist the mirror slightly so that the image falls next to the object. Measure the distance from the lens to the crosswires; this is equal to the focal length of the lens.

13.7 Plane mirrors

You probably already know the **rules of reflection**, which apply to rays of light bouncing off a mirror. Fig. 13.17 reminds you of these rules. If these are new to you, you ought to invent an experiment (using the light rays kit) to test the truth of these rules.

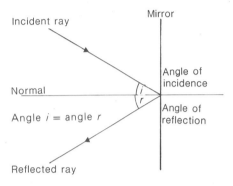

Fig. 13.17 The rules of reflection

Notice that the line at right angles to the mirror is called the **normal** to the mirror. The angle of incidence and the angle of reflection are measured with respect to the normal.

You can see from Fig. 13.18 that the *image* of Beatrice the mouse is as far behind the mirror as she is in front of it. Is the image of Beatrice *real* or *virtual*? (Do the rays of light *really go through the image?*)

Fig. 13.18 Where is Beatrice's image?

The next diagram shows how the camera 'saw' Beatrice in the mirror (the camera could equally well be your eye). Why have some of the lines in the diagram been drawn with dashes?

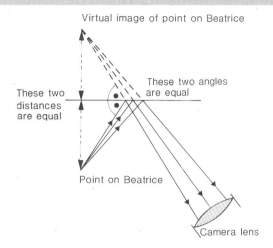

Fig. 13.19 A ray diagram to show how Beatrice's image is formed

13.8 Concave mirrors

Not all mirrors are flat, or plane; curved mirrors have their uses as well. Curved mirrors can be convex, in which the centre curves *towards* you as you look into it, or concave, in which the centre curves *away* from you. This section is about concave mirrors.

Look into a concave mirror, as Kate is doing in Fig. 13.20. Do you obtain an enlarged image of yourself like Kate? Hold the mirror at arm's length from yourself. Does the image change in any way? Does it become larger or smaller, for example?

Fig. 13.20 What do you look like in a concave mirror?

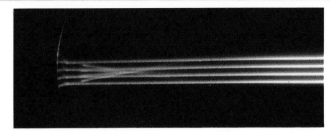

Fig. 13.21(a)
Parallel rays being reflected by a
concave mirror

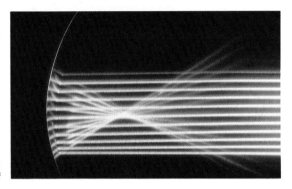

Fig. 13.21(b)
In what way is this photograph
different from the previous one?

Look at Fig. 13.21. This shows parallel rays of light (formed with a light rays kit) striking a concave mirror. As you can see, those in Fig. 13.21(a) are all brought together at one point—the *principal focus* (in a similar way to what sort of lens?) However, as Fig. 13.21(b) shows, rays which hit the mirror near the edge do not pass through the principal focus.

If the mirror were *parabolic* in shape, *all* the rays would be reflected to the principal focus. The reflector of the radio telescope in Fig. 13.22 is parabolic so that it reflects radio waves (which, of course, behave in a similar way to light waves) to a receiving aerial at the principal focus.

The reflector of a car headlamp, or a searchlight, is parabolic. In this case the rays of light *start* at the principal focus (where the lamp filament is) and are reflected in a parallel beam.

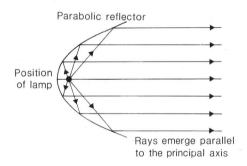

Fig. 13.23   A car headlamp has a parabolic reflector

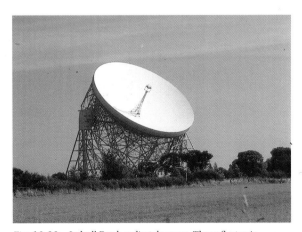

Fig. 13.22   Jodrell Bank radio telescope. The reflector is parabolic in shape

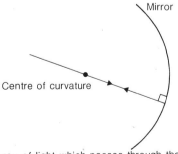

A ray of light which passes through the centre of curvature hits the mirror at a right angle and retraces its path

Fig. 13.24   A ray which passes through the centre of curvature of a concave mirror retraces its steps after reflection

143

If you wish to calculate the position or size of an image formed by a concave mirror, this can be done by drawing an accurate scale diagram in a way similar to that in which you drew a lens. Fig. 13.25 shows how. Notice that, as with a lens, we use two rays of light to locate the position of the image. One is a ray which is travelling parallel to the axis, which is then reflected through the principal focus. The other is a ray which passes through the **centre of curvature** of the mirror. This ray will hit the mirror along a normal (that is, at right angles to the surface of the mirror), and will be reflected back so as to 'retrace its steps'. A diagram like this can, of course, be drawn to scale, preferably using graph paper.

Fig. 13.26   Measuring the focal length of a concave mirror

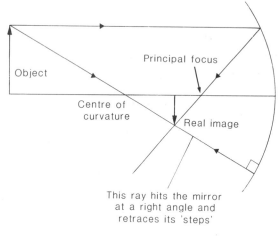

Fig. 13.25   How a concave mirror forms an image

In Fig. 13.25 the object is further from the mirror than the principal focus. Try to draw an accurate ray diagram showing how a concave mirror forms an image of an object which is *nearer* to the mirror than the principal focus. (Hint: you should end up with a virtual image.)

The last photograph in this section shows an experiment to measure the radius of curvature of a concave mirror. Since the radius of curvature is twice the focal length (Fig. 13.28), this experiment also enables you to measure the focal length. An illuminated pair of crosswires are used as the object; the position of the mirror is adjusted so that the image of the crosswires falls on the card next to the object. Rays of light from the point on the object which is on the principal axis have then struck the mirror along the normal (that is, at right angles) and retraced their steps to form an image at the same point (Fig 13.27), so the distance to the crosswires must be equal to the radius of curvature of the mirror.

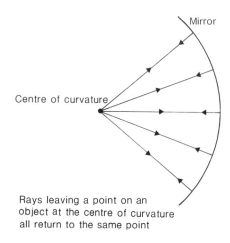

Rays leaving a point on an object at the centre of curvature all return to the same point

Fig. 13.27   A ray diagram for Fig. 13.26

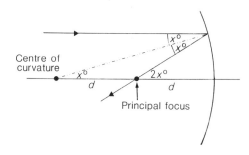

Fig. 13.28   The radius of curvature of a concave mirror is twice its focal length

13.9 Telescopes

Fig. 13.29   The UK 1.2 m Schmidt Telescope is situated on Siding Spring Mountain in Australia

by this lens using a piece of greaseproof paper [Fig. 13.30(b)].

Next mount a strong convex lens (about 15 dioptres) a few centimetres beyond this image. This lens acts as a magnifying glass and magnifies the image made by the objective lens. Remove the paper and look through this **eyepiece lens**; adjust the position slightly until the enlarged image of the lamp or window is sharp. Which way up is the image? Why?

Fig. 13.30(a)   Making a simple astronomical telescope

Fig. 13.30(b)   Stage 2 in making a simple telescope

The telescope in Fig. 13.29 uses a combination of mirrors and lenses and is called a **reflecting telescope**. Does this have any advantage over the first kind of astronomical telescope, which was invented by Johannes Kepler in about 1620 and used two lenses? To be able to answer this question properly, you must understand how each kind of telescope works.

If you have never made a simple astronomical telescope using two lenses (a **refracting telescope**), you should do so now by following the instructions below together with Fig. 13.30.

Mount a weak convex lens (about 2.5 dioptres) on a suitable holder as in Fig. 13.30(a). Point this **objective lens** at a bright object which is some distance away (e.g. a lamp or a window at the far end of the room) and find the real image made

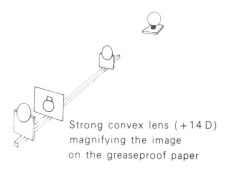

Fig. 13.31   Making a simple astronomical telescope

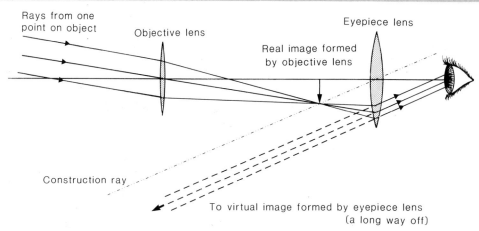

Fig. 13.32   The ray diagram for a simple astronomical
telescope

Fig. 13.32 shows the standard ray diagram for this telescope. Make sure you understand these diagrams. Notice that the image formed by the objective lens is acting as an *object* for the eyepiece lens. If this were an 'ordinary' object we would draw a ray from the end of the object through the middle of the lens and a ray parallel to the axis and through the principal focus in the way you learned in Section 13.3. In this case, the ray through the middle cannot possibly exist; it cannot have come through the objective lens in the right direction. So we draw an 'imaginary' ray, or **construction ray** (shown dashed) to find out the direction of the final virtual image, and then draw in the actual rays afterwards.

It is usual for the eyepiece to be in such a position that the final image is a very long way off (at infinity), as in Fig. 13.32. Can you think of any good reason for this?

Why have the rays coming from *one point* on the object been drawn parallel to each other as they enter the objective lens?

Notice that the eye has been drawn at the position where the rays cross the axis. Why is this the best position for the eye? (Think of how many rays of light would enter the eye if it moved further back, and remember that the diameter of the pupil of the eye is very much less than the diameter of the eyepiece lens.)

An important point to appreciate about a telescope is that it does not necessarily make an image which is bigger than the object, but it makes an image that *appears* bigger, and therefore occupies more of the field of view, as explained in Fig. 13.33. The angle at the eye between the top and bottom of the image is bigger than that between the top and bottom of the object. The telescope produces **angular magnification**.

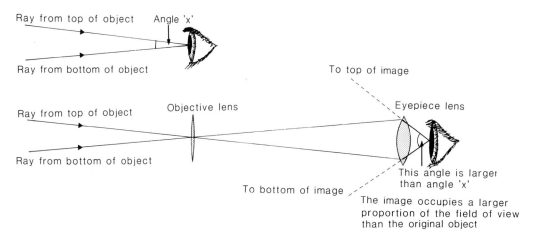

Fig. 13.33   A telescope produces angular magnification

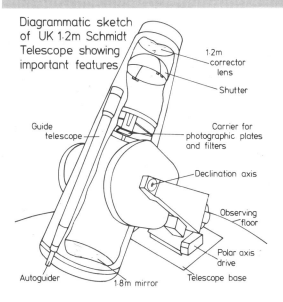

Diagrammatic sketch of UK 1·2m Schmidt Telescope showing important features

- 1·2m corrector lens
- Shutter
- Carrier for photographic plates and filters
- Declination axis
- Observing floor
- Polar axis drive
- Telescope base
- Guide telescope
- Autoguider
- 1·8m mirror

Fig. 13.34   A reflecting telescope—the UK Schmidt Telescope

Fig. 13.35
A photograph of the Orion nebula taken with the UK Schmidt Telescope

The largest telescope of this design is at the Yerkes observatory in Wisconsin, USA. The objective lens is 1 metre in diameter. If an objective lens were made any wider, it would be very difficult to make accurately, and would sag under its own weight. It obviously cannot be supported at its middle because the support would block the light coming through the lens. If a larger telescope is needed it would have to be of a different design.

But why do astronomers need a larger telescope? Many stars are too faint to be seen by your naked eye—not enough light gets into the eye. A telescope with a *wide* objective lens allows more light to enter the telescope and hence it is possible to see fainter stars.

Large telescopes, such as the one illustrated at the start of this section, use an objective *mirror* rather than a lens to form the first (real) image. A mirror can be supported underneath to prevent sagging. A diagram of this telescope is in Fig. 13.34. It is used exclusively for making photographs (such as the one in Fig. 13.35), so a photographic film can be placed at the focal point, as shown. This only blocks out a very small proportion of the rays entering the telescope, and the image will still be complete, even with these rays missing.

The next diagram shows how the rays are often reflected out of the side of the telescope so that an astronomer can look at the stars. This arrangement was used by Newton, who invented the reflecting telescope.

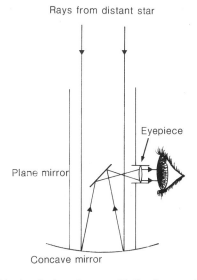

Rays from distant star

Eyepiece

Plane mirror

Concave mirror

Fig. 13.36   A reflecting telescope with the observer at one side

13.10   The compound microscope

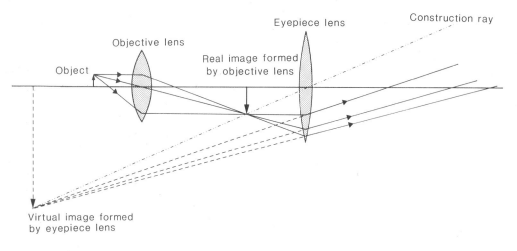

Fig. 13.37   A ray diagram for a microscope

The first compound microscope was made in 1590. The construction of a compound microscope is very similar to that of a refracting telescope, with two convex lenses. In the case of a microscope, however, the objective lens is much stronger than that of the eye lens, as shown in Fig. 13.37. Remember that a microscope objective lens is very close to what is being looked at and the focal length of the objective lens must be *less* than the distance between lens and object. Why?

As with the telescope, the objective lens forms a real image which then acts as an object for the eye lens. In Fig. 13.37 the final virtual image has been drawn fairly close to the eye lens. In what way is this different from the way in which the telescope was drawn?

A microscope in 'normal adjustment' has the image at the **near point** of the eye, i.e. the nearest point on which the eye can focus—usually taken as 25 cm for an adult eye. (Where is *your* near point?) The microscope does not *have* to be used in this way—the final virtual image can be a long way off, as in the telescope. This is more relaxing for your eye, but there is an advantage in having an image close to your eye when you are having to *draw* what you see. Try to think what this advantage is.

13.11   Bending light

In the last chapter you were reminded that waves can diffract, interfere, reflect and refract. You have now seen examples of light waves doing the first three of these, but have you ever seen light *refract*,

that is, bend and move in a different direction? The answer must be yes. Lenses change the direction of rays of light, and you have spent some time studying lenses.

In Fig. 13.38 you can see a ray of light entering a tank of water. Notice how the light bends, or refracts, only when it goes from one substance to another, such as from air to water. As shown in the next diagram, the light bends *towards* the normal (the line at right angles to the surface) as it enters the water. Light travels more slowly in water than in air; look back at Fig. 12.4 if you cannot remember why this should make the light bend in the way it does. The light bends the opposite way as it enters the air again. Notice that the light emerges from this tank of water parallel to its original direction.

Fig. 13.38   A ray of light being refracted as it passes from air to water. Notice that there is also some reflection

In 1620, a Dutch scientist called Snell discovered a relationship between the angle of incidence i and the angle of refraction r. For light passing from one medium to another, the value of $\sin i/\sin r$ is constant. (This is known as **Snell's law**.)

If light is passing from a vacuum to a substance, the value of $\sin i/\sin r$ is called the **refractive index** of the substance. For most practical purposes, it makes no difference if the light goes from air to the substance rather than from a vacuum.

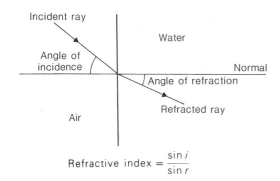

$$\text{Refractive index} = \frac{\sin i}{\sin r}$$

Fig. 13.39

The refractive index is related to the *speed* of light in the substance as follows:

$$\text{refractive index} = \frac{\text{speed of light in vacuum}}{\text{speed of light in substance}}$$

Design an experiment, preferably using the light rays kit and, say, a tank of water as in Fig. 13.38 to test Snell's law for light travelling from air to water. You will have to think of a sensible way of measuring the angles i and r before using a calculator to find the values of $\sin i$ and $\sin r$.

Since

$$\frac{\sin i}{\sin r} = \text{constant}$$

$$\sin i = \sin r \times \text{const}$$

so a graph of $\sin i$ against $\sin r$ should be a straight line. Why? Do your results give a straight line graph when you plot $\sin i$ against $\sin r$? What does the gradient of your graph tell you?

Refraction is responsible for the odd effect in Fig. 13.40, where the depth of the glass block appears to be less than it really is. Fig. 13.41 explains why. Have you ever noticed this effect anywhere else?

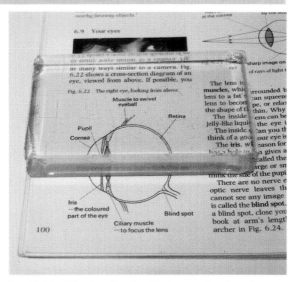

Fig. 13.40 Why does this glass block appear less deep than it really is?

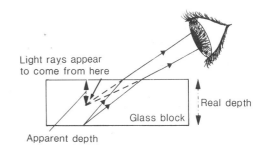

Fig. 13.41 Explaining Fig. 13.40

Refraction is also responsible for that familiar desert illusion, the mirage. A mirage can sometimes be seen just above a hot road surface on a summer's day. Fig. 13.42 explains how a mirage arises. The air just above the road surface, or desert, is hotter than normal, so the refractive index is less than the cooler air above. Rays coming from the sky, or from objects along the road, are gradually bent away from the normal. As you can see in the diagram, an observer sees a virtual image which looks like a reflection in a pool of water.

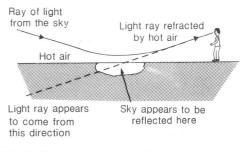

Fig. 13.42 How mirages are formed

149

13.12   Fibre optics

Fig. 13.43   Laying an optical fibre cable

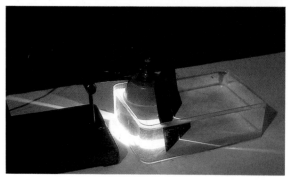

Fig. 13.44   This apparatus can be used to investigate what happens to rays of light leaving a tank of water

You can see that the angle of incidence in Fig. 13.45(b) is greater than in Fig. 13.45(a). Does it look as if the amount of *reflected* light is the same in both cases? If not, in which case is more light reflected?

a)

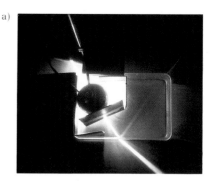

b)

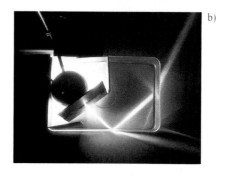

c)

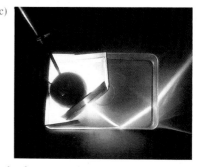

Fig. 13.45   Notice what happens to the light as the angle of incidence increases

The men in this photograph seem to be doing a perfectly normal job—laying a telephone cable. The slightly unusual thing about the cable is that it is made of glass. This may seem very silly to you; glass does not conduct electricity, and we all know that telephone conversations are carried by electricity. But this cable is not meant to carry electricity, it carries light, and scientists have found a way of using light to carry telephone conversations, as well as television signals, data between computers and so on. A large proportion of the telephone conversations between London and Birmingham, for example, is carried in glass cables.

The glass in the cable is in the form of very thin fibres. The glass from which the fibres are made must be very pure, or the light will be absorbed. But how is the light guided along these fibres?

Look at Fig. 13.38 again. Notice that, not only is the ray of light *refracted* at each surface, but some of the light is also *reflected*. This always happens when light passes from one substance to another.

Why do you think this effect is a problem for the designer of a camera lens, where there may be six or more separate 'elements', or lenses, to form one complete lens system?

Now look at Fig. 13.44. This shows a similar arrangement to Fig. 13.38, but the lamp has been placed *inside* the tank of water, so that the rays are only going from water into air. If you try this experiment for yourself, put the lamp in the water before you switch on the lamp; if you switch on the lamp first you may break the bulb as the hot bulb touches the cold water.

Notice that the angle of refraction is *greater* than the angle of incidence, since the light is travelling into an optically less dense medium, that is, *from* glass *into* air. As we make the angle of incidence larger, there comes an angle where *all* the light is just reflected internally—Fig. 13.45(c). This is called **total internal reflection**. The angle of incidence in this case is called the **critical angle**.

If the angle of incidence is greater than the critical angle, then all the light will still be totally internally reflected.

This is exactly what is happening to the light inside an optical fibre—it is being totally internally reflected. One kind of optical fibre consists of two very thin glass rods, one inside the other; the inner rod has a slightly higher refractive index than the outer layer. Light travelling along the central core is totally internally reflected off the outer core, as in the diagram, and so zig-zags its way down the fibre.

The whole fibre is coated with a protective layer of plastic to prevent it becoming scratched.

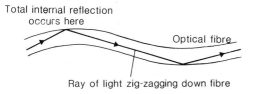

Fig. 13.46   Rays travelling down an optical fibre cable

Why bother with optical fibres? One of the most important and significant results of the microelectronics revolution is that the amount of information in the form of television, telephone conversations, data between computers and so on, that needs to be passed around the country, and the world, is increasing rapidly. Communications engineers are faced with an ever growing demand for better communications links. Optical fibres have several advantages over conventional copper cables. They are very light: 500 m of fibre cable has a mass of 25 kg; the same length of copper co-axial cable has a mass of 5 tonnes. More important, light can carry much more information at a time than electric signals; the fibre illustrated in Fig. 13.43 can carry over 2000 telephone conversations simultaneously.

Optical fibres are free from interference caused by electrical machinery, radio waves or lightning, for example. This makes them attractive for carrying information between parts of a computer.

Probably the biggest advantage is their ability to carry a signal for long distances without amplification. Amplifiers, or 'boosters', are needed every 2 km with a conventional cable, to compensate for loss of signal strength as a result of the electrical resistance of the wire. As this book is being written (1984), the longest distance that an optical fibre has carried a signal without amplification is over 100 km, and engineers are seriously talking of a transatlantic optical fibre cable without amplifiers. Such is the rate of progress in this field that it is quite probable that such a link between Europe and America will exist by the time you read this book.

The idea of piping light along glass is not very new. Optical fibres have been used in medicine, for example, to pipe light to places which are normally inaccessible, to illuminate various organs inside a patient's body so that they can be observed using suitable instruments. The 'light pipe' consists of many optical fibres in a bundle. Another way of using light pipes is to illuminate many things using just one lamp, which means there is just one lamp to maintain.

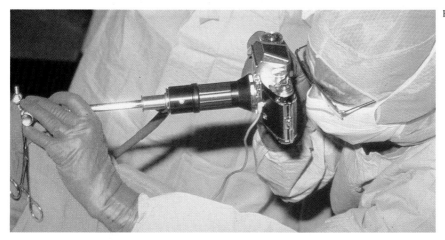

Fig. 13.47   This surgeon is using an optical fibre endoscope to examine a patient internally

13.13 More total internal reflection

Total internal reflection has other uses apart from in fibre optics. Total internal reflection is taking place in Fig. 13.48. The ray is reflected twice in the prism before emerging parallel to its original direction.

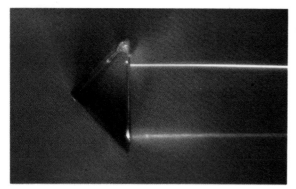

Fig. 13.48 A ray being totally internally reflected twice in a prism

Draw a diagram showing the path of the ray of light inside the prism. The critical angle for the glass that the prism is made of must be less than 45°. Explain why.

Prisms are used in this way in binoculars to 'fold up' the path of the rays of light so that the binoculars can be relatively short (see Fig. 13.49).

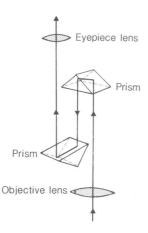

Fig. 13.49 Binoculars

Some cameras (single lens reflex or SLR cameras) use a special five-sided prism (a pentaprism) in the viewfinder system, so that the user can look through the camera's lens and see exactly what appears on the film. Fig. 13.50 shows how the prism is used. Why use a *five-sided* prism? Wouldn't an ordinary three-sided prism be satisfactory? (You will have to think *very* hard about this one!)

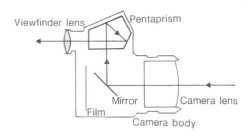

Fig. 13.50 A pentaprism in the viewing system of a single lens reflex camera

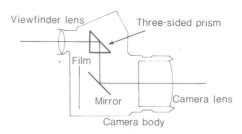

Fig. 13.51 Why will this system not do instead of a pentaprism?

SUMMARY

Now that you have finished studying this chapter on lenses, mirrors and rays of light, there are a number of things you should know and be able to do.

1 You should:
   a) be able to draw ray diagrams for
      i) a convex lens forming a real image
      ii) a convex lens forming a virtual image (magnifying glass)
      iii) a concave mirror forming a real image
      iv) a concave mirror forming a virtual image;
   b) be able to calculate the size and position of an image using a scale diagram;
   c) be able to perform an experiment to determine the focal length of a convex lens;
   d) be able to perform an experiment to determine the focal length of a concave mirror;
   e) know that for a mirror, the radius of curvature is twice the focal length;
   f) know the rules of reflection;
   g) be able to explain how a simple refracting astronomical telescope works;
   h) understand how a reflecting telescope works, and what advantages it has over a refracting telescope;
   i) be able to explain how a microscope works;
   j) know the laws of refraction;
   k) be able to explain how a mirage occurs;
   l) know what is meant by total internal reflection, and how it occurs;
   m) understand how an optical fibre can carry light;
   n) know a use of fibre optics;
   o) know at least one other use of total internal reflection.

2 You should know what each of the following is, or what it does:

principal axis	real image
virtual image	normal
principal focus	parabolic mirror
centre of curvature (of mirror)	objective lens
	construction ray
critical angle	refractive index
eye lens	

3 These are some of the other words that have been used in this chapter. You should know what each word means:

construct	basis
crosswires	exclusively
significant	information
revolution	communications
conventional	compensate
inaccessible	

FURTHER QUESTIONS

1 If you look out of a window during the day, you see whatever is outside the window. If you look out of the same window at night, with the lights on in the room, you tend to see whatever is *inside* the room. Explain why this is so.

2 Why do you sometimes see signs like the one illustrated in Fig. 13.52?

ƎƆИAⅬUꓭMA

Fig. 13.52

3 Explain, using diagrams to help your explanation:
   a) why a swimming pool appears shallower than it really is;
   b) why the oars of a rowing boat seem broken as they enter the water;
   c) why if a diver who is underwater looks up at the surface of the water, most of the surface seems silvery;
   d) why bubbles of air from a diver look silvery.

4 Explain how a fishbowl of water could start a fire.

5 A slide projector has a lens of focal length 10 cm. A slide is placed in the projector and an image of the slide is focused onto a screen 1 m away. By drawing a suitable scale diagram, find:
   a) the distance between the slide and the lens;

b) the width of the image on the screen if the slide is 35 mm wide.

6 A boy looks into a concave mirror of focal length 30 cm and sees an image of his own face. He holds the mirror 20 cm from his face. By drawing a suitable scale diagram, find:
a) where the image of his face is;
b) which way up the image is;
c) how much bigger or small the image of his face is compared to his actual face.

7 Fig. 13.53 shows a convex lens placed 50 cm from a small bright object O. An image of this object is formed at position I, 25 cm from the lens.

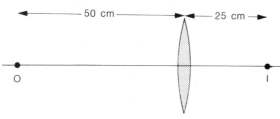

Fig. 13.53

a) If the object is moved 4 cm below its present position, where will be the new image position?

b) Draw a careful diagram to show how the lens forms an image of the object in its new position. Draw the paths of at least two rays from the object to the image.
c) If the object now moves away from the lens (starting 4 cm below its original position), what happens to the image?

8 Fig. 13.54 shows a ray of light zig-zagging along an optical fibre which is slightly curved.

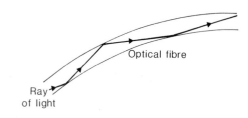

Fig. 13.54

a) Explain why the ray of light stays inside the optical fibre, rather than coming out through the surface of the fibre.
b) Explain why an optical fibre cannot be bent into too sharp a curve.

ELECTRICAL ENERGY

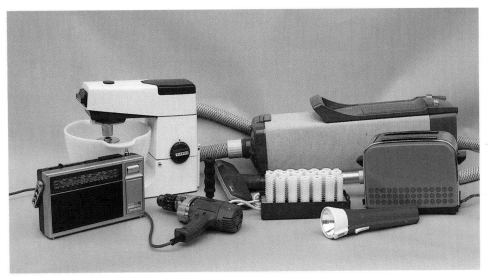

Fig. 14.1   All these items use electrical energy

14.1   Introduction

Of all the kinds of energy which you use, you will surely agree that electrical energy is the most convenient. It is usually available at the flick of a switch, and you probably just take it for granted. As you learned in Chapter 1, a large proportion of our fossil fuels are converted to electrical energy.

Wouldn't your life be much more difficult without electricity? The photograph shows some of the everyday objects which use electrical energy. Would you like to be without them? Could you manage without them? What other things are there which use electrical energy and which you would not like to be without?

14.2   You should know ...

Before you can understand this chapter properly, you must know:
1 how to draw electric circuit diagrams;
2 how to use an ammeter to measure electric current;
3 how series and parallel circuits behave;
4 what is meant by electrical resistance;
5 that electric current cannot 'get lost'; it is conserved.

In case you have forgotten or are not sure about these points, here is a quick reminder of some of them, but for further details you must look in another book such as *Physics in Action*.

Fig. 14.2 shows the **circuit symbols** that you should know. The older symbol for a resistor is also shown since you might see it in older text books

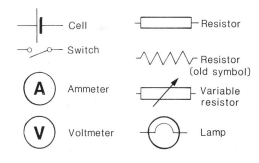

Fig. 14.2   Symbols used in electric circuit diagrams

and want to know what it means. These symbols are used to draw circuit diagrams like the one in Fig. 14.3.

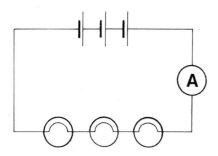

Fig. 14.3   A circuit diagram showing three bulbs in series

Fig. 14.3 includes an ammeter to measure the current flowing in the circuit. The next photograph shows two versions of this circuit, one using an analogue ammeter, and the other using a digital ammeter. Notice how the ammeter is connected into the circuit, with all the current you wish to measure passing through the ammeter. With an analogue meter you often have to put the decimal point in the right place on the scale according to

which shunt is plugged in. In Fig. 14.4(a) the ammeter reads 0.3 A, (not 3 A). A digital meter usually puts the decimal point in the right place for you. You must make sure you can correctly read the scale on the ammeters you use.

The lamps in the circuit illustrated in Fig. 14.3 are **in series**. The same current passes through all three lamps; it does not matter where in the circuit you put the ammeter, it will read the same current.

If *more* lamps are put into this circuit, it becomes more 'difficult' for the electric current to pass round the circuit, in other words, the **resistance** of the circuit increases, with the result that the current goes down, assuming the same number of cells are used. The *higher* the resistance, the *smaller* the current.

In Fig. 14.5 the lamps are **in parallel**. There are two alternative routes along which the electricity can flow on its journey from one side of the cell to the other. The same current will flow along each route only if each route is of the same resistance, which is the case in Fig. 14.5 provided the two lamps are identical. The current flowing out of the cell will be the same as the current flowing back in, as can be seen in Fig. 14.6; current cannot 'get lost'.

Fig. 14.4(a)   The circuit illustrated in Fig. 14.3, using an analogue ammeter

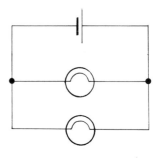

Fig. 14.5   The bulbs in this circuit are in parallel

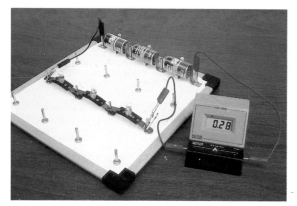

Fig. 14.4(b)   The same circuit with a digital ammeter

Fig. 14.6   Measuring the current leaving and returning to the battery

If another lamp is added parallel to the first two, this provides another route along which electricity can flow, so *more* electricity will flow from the cell, as illustrated in Fig. 14.7.

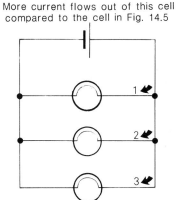

More current flows out of this cell compared to the cell in Fig. 14.5

There are three alternative routes by which electricity can flow round this circuit

Fig. 14.7 Three parallel bulbs

14.3 Electric charge

Fig. 14.8

The rate at which water is flowing under this bridge might be measured in gallons per minute or, to use more scientific units, litres per second. This 'rate of flow' is equivalent to the *current* in an electric circuit. You can also measure a *quantity* of water, such as a bucketful, or a litre, or a cubic metre.

Is there an equivalent thing for electricity? Can you talk about a 'quantity of electricity'? The answer is yes, except we usually refer to a 'quantity of electric charge'. Although you probably think of electric charge as something which flows along a wire in the same way as the water is flowing under the bridge, you *can* collect small quantities of electric charge in a 'bucket' just as you can hold water in a bucket. You will learn more about this in Chapter 19 which is about static electricity.

Electric charge is measured in **coulombs**; a coulomb of charge flows if a current of 1 ampere flows for 1 second. The equation that relates electric charge and electric current is therefore:

$$\text{electric charge} = \text{electric current} \times \text{time}$$

or, in terms of units:

$$\text{coulombs} = \text{amperes} \times \text{seconds}$$

What is the total charge which flows if a current of 0.5 ampere flows for 30 seconds?

What is the electric current flowing if a total charge of 50 coulombs flows in 10 seconds?

14.4 Measuring electrical energy

Look at Fig. 14.9. On the left, a single cell is lighting a torch bulb; on the right, a light bulb is being lit from the mains. In which circuit is electrical energy being converted into heat and light energy at the greater rate? Can you guess which circuit has the greater current flowing in it?

Fig. 14.9 Which bulb has the greater power?

(*CAUTION This experiment is dangerous since it involves mains electricity. Your teacher will probably want to demonstrate it to you.*)

Now look at Fig. 14.10, which shows the same circuits with ammeters added so that the current can be measured. Both ammeters show the *same* current. Does this surprise you?

Fig. 14.10   Measuring the current to the two bulbs in the previous photograph

How are you going to explain this? There is clearly much more energy being converted per second in the 'mains' bulb. However, there is exactly the same number of coulombs flowing through the mains bulb every second as through the torch bulb. To help you understand what is happening, the apparatus illustrated in Fig. 14.11 is useful. You have probably seen this apparatus before. It is called a water circuit board. The pump at the bottom pumps the water (which has been coloured so that you can see it) round the pipes. Notice that the pump is not *making* water; the water is already in the pipes. The pump is giving it energy so that it can get round the water circuit, especially through the narrow pipes at the top of the circuit.

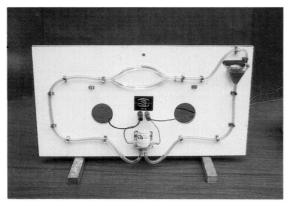

Fig. 14.11   A water circuit board

An electric circuit behaves in a similar way. A cell does not *make* electric charge, the charge is already in the wires; the cell gives each coulomb of charge some electrical energy, which is converted into other forms of energy as the charge travels round the circuit. The generator at a power station does a similar job to a cell.

Scientists say that the water circuit is an *analogy* for an electric circuit. Of course, it is not absolutely identical—after all, water is not electricity! Why do you think scientists use analogies?

We can explain the behaviour of the circuits in Fig. 14.10 if we assume that the mains gives each coulomb *more* energy than a single cell. Then each coulomb passing through the mains bulb carries more energy than can be converted to heat and light (see Fig. 14.12).

Coulombs of electric charge leave this cell with a low quantity of energy

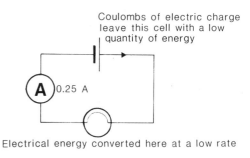

Electrical energy converted here at a low rate

Coulombs of electric charge leave the mains with a large quantity of energy

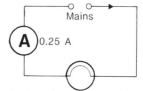

Electrical energy converted here at a high rate

Fig. 14.12   The electric charge from the mains has more energy than the charge from a battery

What would be useful is an instrument that will plug into an electric circuit to *measure* the energy carried by each coulomb. Unfortunately, this is not directly possible, but there is an instrument that can measure the *difference* between the energy carried by each coulomb in one place in a circuit and the energy carried by each coulomb at another place in the circuit. This instrument just plugs into the two points at which you wish to compare energies as illustrated in Fig. 14.13. The proper

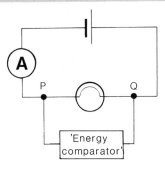

This shows the difference between
the energy of each coulomb at P
and the energy of each coulomb at Q

Fig. 14.13   A 'voltage comparator'

name for this instrument is a **voltmeter**, and the next photograph shows a voltmeter connected across each of the bulbs illustrated in Fig. 14.9. The reading of 1.25 V on the left-hand meter means that each coulomb of electricity passing through the torch bulb loses 1.25 J of energy; the reading of 240 V on the right-hand meter means that each coulomb of electricity passing through the large bulb loses 240 J of energy (see Fig. 14.15).

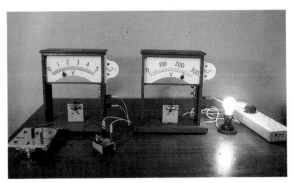

Fig. 14.14   A voltmeter in use with the circuits in Fig. 14.9

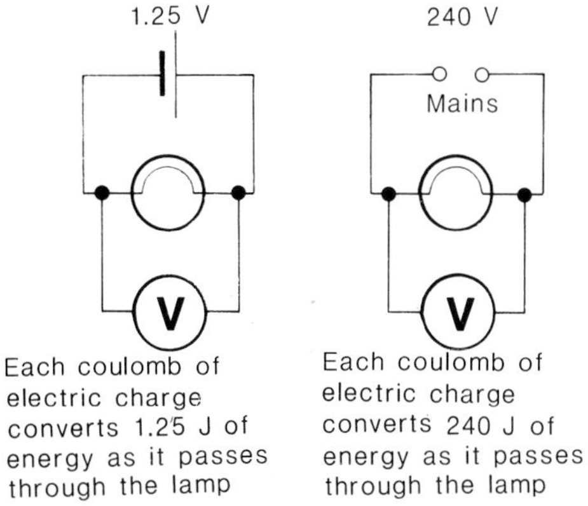

The correct wording is to say that the **potential difference** across the torch bulb is 1.25 volts; the potential difference across the large bulb is 240 volts. Notice the word 'difference' being used to remind you that we are referring to a *difference* in energies.

A formal definition of a volt is as follows:

If a potential difference of 1 volt exists between two points in a circuit, then for each coulomb of charge travelling between these two points, 1 joule of energy is converted.

Notice that it does not matter whether the coulomb *gains* energy (as it would as it passes through a cell) or *loses* energy, as it would in a light bulb, for example.

An easier, but less precise, definition to remember is:

'one volt = 1 joule per coulomb'.

From this definition comes the following useful equation:

$$\text{potential difference} = \frac{\text{electrical energy}}{\text{charge}}$$

or, in units:

$$\text{volts} = \frac{\text{joules}}{\text{coulombs}}$$

14.5   Using a voltmeter

You have just learned that a voltmeter measures the change in energy carried by every coulomb of charge passing between two points in an electric circuit, and that the voltmeter must be connected *between* these two points, as in Fig. 14.16.

1.25 V

Each coulomb of electric charge converts 1.25 J of energy as it passes through the lamp

240 V

Mains

Each coulomb of electric charge converts 240 J of energy as it passes through the lamp

Fig. 14.15

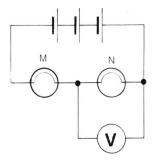

Fig. 14.16   A circuit diagram which includes a voltmeter

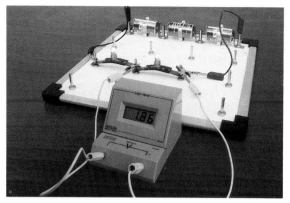

Fig. 14.17   Connecting a voltmeter in an electric circuit

When you are setting up an electric circuit containing a voltmeter, it will help you to do so correctly if you connect the voltmeter last; you should be able to connect it without disconnecting any wires or disturbing the rest of the circuit in any way. To practise using a voltmeter, set up the circuit in Fig. 14.16, then answer the questions.

1 What is the potential difference (voltage) across lamp M?
2 Explain what this means, using the words 'joules' and 'coulombs'.
3 What is the potential difference across lamp N?
4 What is the potential difference across both lamps together?
5 What is the potential difference across each cell?
6 What is the potential difference across all three cells together?
7 Explain exactly what this means in terms of energy (in joules) and electric charge (in coulombs).
8 Compare your answer to Question 4 with your answer to Question 6. Have you any comments to make?

14.6   A numerical example

Fig. 14.18 shows five small lamps, rated at 3 watts, joined in series with one 100 watt lamp to the mains. The ammeter reads 0.5 A.

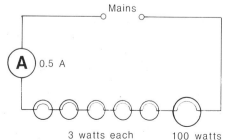

Fig. 14.18

1 How many coulombs pass through each small lamp in 1 second?
   Answer   0.5 coulomb. A reading of 0.5 A means 0.5 coulomb every second.
2 How many joules of electrical energy are converted to heat and light energy in each small lamp in 1 second?
   Answer   3 joules every second. 3 watts means 3 joules every second.
3 How many joules of electrical energy are converted by each coulomb passing through a small lamp?
   Answer   6 joules. In 1 second, 0.5 coulomb passes through a lamp, converting 3 joules of energy, so 6 joules of energy will be converted by one whole coulomb.
4 What is the potential difference across each small lamp?
   Answer   6 volts, since there are 6 joules of energy being converted by each coulomb (1 volt is 1 joule per coulomb).
5 How many coulombs pass through the large lamp in 1 second?
   Answer   0.5 coulomb. Why?
6 How many joules of electrical energy are converted in the large lamp in 1 second.
   Answer   100 joules in 1 second. Why?
7 How much energy does each coulomb convert in the larger lamp?
   Answer   200 joules for every coulomb.
8 What is the potential difference across the larger lamp?
   Answer   200 volts.
9 How much energy must the mains supply to each coulomb of electric charge?
   Answer   230 joules for every coulomb. 200 of these joules will be converted in the large lamp, and 6 joules in each of the 5 small lamps.
10 What is the potential difference of the mains?
   Answer   230 V. Why?

14.7   Electric power

Earlier in this book, in Chapter 8, you learned that engineers and physicists are often more interested in *power* than in energy. Why is this? Look back at Chapter 8 if you cannot remember.

How do we measure electric power, that is, electrical energy transferred every second? Let us start with the definition of power:

$$power = \frac{energy}{time}$$

In the case of electricity, you know that

energy = potential difference × charge

therefore:

$$\text{electric power} = \frac{\text{potential difference} \times \text{charge}}{\text{time}}$$

But

charge = current × time

so

electric power = potential difference × current

or, in terms of units:

watts = volts × amperes

This is a very important equation.

Now you should know the following two important statements:

1 Energy: joules = volts × coulombs
2 Power: watts = volts × amperes

In order to measure the electric power of any device, you must measure the current *through* the device and the potential difference *across* the device. For example, to measure the power of a torch bulb, the circuit shown in Figs 14.19 and 14.20 is needed. Notice how the voltmeter is connected across the bulb.

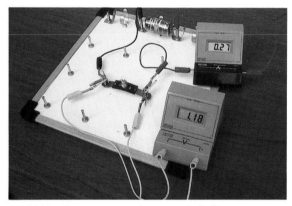

Fig. 14.19   How to measure the power of a bulb

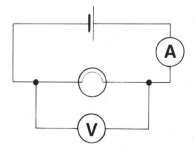

Fig. 14.20   The circuit diagram for Fig. 14.19

What is the potential difference across the bulb in the photograph? What is the current flowing in the circuit? What is the power of the torch bulb (volts × amperes)? Remember to include the correct units of power.

Practise setting up the correct circuits and making measurements of power for yourself by measuring the power of suitable electrical devices in your laboratory, for example, the car light bulb in Fig. 14.21, or the immersion heater you used in Section 9.5.

Fig. 14.21   How would you measure the power of this bulb?

14.8   Measuring the power and efficiency of an electric motor

Write down what we mean by the efficiency of a machine. (If you have forgotten, you first met this idea in Chapter 8.) In order to *measure* the efficiency of an electric motor, you need to measure the electric power you *supply* to the motor, and the useful power you obtain *from* the motor (the 'output power'). If you did an experiment to measure the output power of an electric motor in Section 8.4, now that you know how to measure the electric power supplied to the motor, you can measure how good your motor is at converting electrical energy to gravitational potential energy (in other words, measure the efficiency of the motor).

A suitable apparatus is shown in Fig. 14.22. It is very similar to that shown in Fig. 8.8, but now includes a voltmeter and ammeter in the electric circuit so that the electric power can be measured.

Fig. 14.22　Measuring the power of an electric motor

Assemble this apparatus for yourself if you can. Tie a load onto the string (about 5 N is suitable for the motor shown in the photograph; the weight should be such that, when the motor is switched on, the load rises at a constant rate). Switch on the motor and take the following readings, recording your results in a suitable table like the one below.

Table of results	
Potential difference across motor	V
Current through motor	A
Weight of load being lifted	N
Time taken to lift load	s
Distance load is lifted	m

1　Which two readings do you need to calculate the input power to your motor?
2　What do you do to these two readings to find the input power?
3　What was the input power to your motor?
4　How much potential energy did the load gain?
5　Why do you need to know the time taken to lift the load?
6　What was the output power of your motor?
7　Why is the output power less than the electric power you put into the motor?
8　What was the efficiency of your motor?

14.9　Electricity in your home

Fig. 14.23　What is wrong here?

What a horror (Fig. 14.23)! Such an arrangement is downright dangerous—but why? There are no frayed wires; the plugs are pushed firmly in. What is wrong with this arrangement?

To answer this question you need to know a little about how the electric circuits are arranged in your house.

Fig. 14.24 shows a typical arrangement for the electric wiring in a house built in the last 30 years or so, and in older houses that have been rewired. Electricity is supplied to a **distribution box**, which is usually combined with a **fuse box** or **consumer unit**. From there, circuits (usually two of them) go round part of the house, supplying each socket in turn, finally returning to the fuse box, so forming a **ring main**. Electricity can go either way round a ring to any socket. (Older houses used to have individual wires going from the fuse box to each individual socket; this used much more wire than a ring main system, so was less economical.)

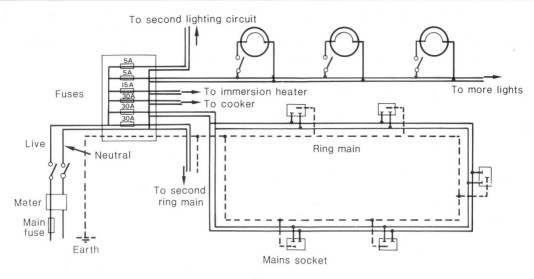

Fig. 14.24   A typical arrangement for wiring in a house

Fig. 14.25   A distribution box in a house

A ring main is usually designed to carry a maximum current of 30 A, so there is a 30 A fuse in the fuse box to prevent more current than this flowing (Fig. 14.26). What happens to the fuse in the photograph if more than 30 A flows through it? Some consumer units have **circuit breakers** instead of fuses. These are special switches which switch off the circuit if too much current is flowing. Do you think they have any advantages compared with traditional fuses?

Electric cookers and immersion heaters are not plugged into a ring main; they have their own separate cable straight from the fuse box. The cable to the cooker is capable of carrying 30 A. If the mains voltage is 240 V, the maximum power that can be delivered to the cooker is 240 V × 30 A = 7200 W (or 7.2 kW).

Fig. 14.26   Removing a 30 A ring mains fuse

An immersion heater is usually wired with a 15 A cable. What is the maximum possible power of an immersion heater?

Suppose, by mistake, a cooker was wired with a 15 A cable. What do you think would happen to the cable when the cooker was switched fully on?

163

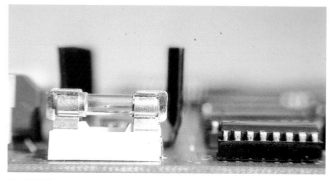

Fig. 14.27 How much current does this 3 kW kettle take from the mains?

Fig. 14.29 A piece of electronic equipment protected by a separate fuse

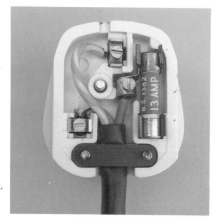

Fig. 14.28
Inside a mains plug. Notice the fuse on the right

Fig. 14.30 What current passes through this 60 W, 240 V mains lamp?

The kettle plugged into the ring mains (Fig. 14.27) is rated at 3 kW. How much current will it draw from the 240 V mains?

Using $P = V \times I$

(What do these letters stand for?)

$$3000\,\text{W} = 240\,\text{V} \times I$$

$$I = \frac{3000\,\text{W}}{240\,\text{V}}$$

$$I = 12.5\,\text{A}$$

The current drawn by this kettle is 12.5 A.

Suppose something went wrong with the kettle so that a much larger current started to flow—say 25 A? Clearly the plug, the socket and the cable leading to the kettle will become hot unless the current is quickly stopped. The 30 A fuse at the fuse box will not help—why not?

To deal with this sort of problem, the plug has a fuse in it that will melt if too much current flows through the kettle. You can see this fuse in Fig. 14.28; it is rated at 13 A. This is one of the two common fuse ratings found in the home, the other being 3 A.

It is possible to obtain fuses rated at different values, but these are not common in the home. Some appliances, for example, some televisions, hi-fi equipment, as well as much of the equipment you use in the lab, also have a fuse in the equipment itself.

It is important to fit the correct fuse to a plug. Suppose the lamp in Fig. 14.30 had a 13 A fuse fitted to its plug. The lamp has a 60 W bulb. Assuming the mains voltage is 240 V, the current through this lamp is 0.25 A. (How was this calculated?) The current would have to rise to 52 times its normal value before the fuse would blow. A 3 A fuse would be the right choice here.

Something which draws 3 A from the mains will have a power of 240 V × 3 A = 720 W. Therefore anything with a power rating of less than about 700 W should be fitted with a 3 A fuse, and anything more powerful should have a 13 A fuse.

Now let us return to the horror in Fig. 14.23. The kettle is rated at 3 kW, the toaster at 1 kW and the iron at 800 W. The total power being drawn is 4800 W, so the current being drawn is given by

$$\text{current} = \frac{\text{power}}{\text{p.d.}}$$

$$= \frac{4800\,\text{W}}{240\,\text{V}}$$

$$= 20\,\text{A}$$

This is not enough current to blow the fuse at the fuse box. None of the individual fuses in the plugs will blow (Why not?), but 20 A is far more current than the socket and adaptors are designed to carry. What is likely to happen to them?

If, at any time, it is essential to plug more than one thing into a mains socket, the distribution board illustrated in Fig. 14.31 is a much safer thing to use since the board's plug is fitted with a 13 A fuse. What would happen if all the appliances illustrated in Fig. 14.23 were plugged into this distribution board? Why is this board safer than two- or three-way adaptors?

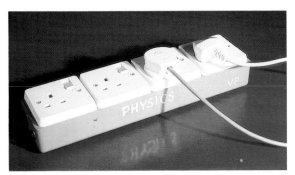

Fig. 14.31   A four-way distribution board in use

14.10   Fusing and earthing

Look again at Fig. 14.28. Why are there *three* wires and three pins on the plug? The brown wire is the **live** wire, which must be connected to the fused terminal on the plug. The blue wire is the **neutral** wire along which electricity returns to the mains. What is the third, green and yellow, wire for?

It is called an **earth** wire; it connects the outside of the kettle to the ground outside your house via the top pin on the plug, as shown in Fig. 14.32. It is an essential safety feature; all mains appliances with a metal casing should be earthed in this way. Why?

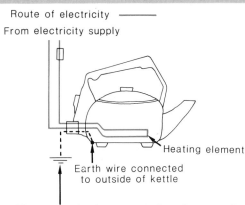

Route of electricity ——————
From electricity supply

Heating element

Earth wire connected
to outside of kettle

The earth wire is connected to the ground
via the plug and the house wiring

Fig. 14.32   'Earthing' a kettle

Suppose a fault occurs in a kettle which was not earthed and the live wire touches the outside of the kettle (Fig. 14.33). If anybody touches the kettle, electricity will flow through that person (to the earth) and electrocute them. (The Earth can act as a large 'reservoir' of electricity.) If the outside of the kettle is earthed, as soon as a fault develops a large current will flow along the earth wire to the ground, and this current should blow the fuse in the plug, so disconnecting the kettle from the mains and making it safe to touch.

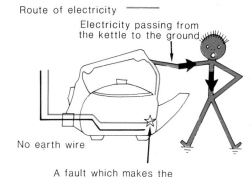

Route of electricity ——————

Electricity passing from
the kettle to the ground

No earth wire

A fault which makes the
outside of the kettle live

Fig. 14.33   A fault in the wiring of a non-earthed kettle could be lethal

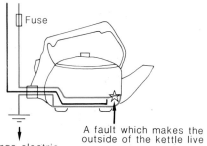

Route of electricity

Fuse

A fault which makes the
outside of the kettle live

A large electric
current flows to earth.
This will blow the fuse

Fig. 14.34   How earthing makes the faulty kettle safe

14.11 Paying for electricity

You have to pay the Electricity Board for the electrical energy you use. Although you measure energy in joules, the Electricity Boards find this unit of energy too small, and they use a unit called a **kilowatt hour** (kW h). This is the energy converted in 1 hour by an appliance whose power is 1 kilowatt.

Since a 1 kW appliance converts 1000 J of energy in 1 second, in 1 hour it will convert $1000 \times 3600 = 3\,600\,000$ J of energy. (There are 3600 seconds in an hour.) Therefore, 1 kW h is equal to 3 600 000 J of energy.

There will be a meter somewhere in your house which records the number of kilowatt hours of energy that have been supplied. The Electricity Board reads this meter, usually every three months, and sends a bill (Fig. 14.36) for the amount of energy used. As you can see, electrical energy cost 5.10p per 'unit' (kilowatt hour) in March 1983.

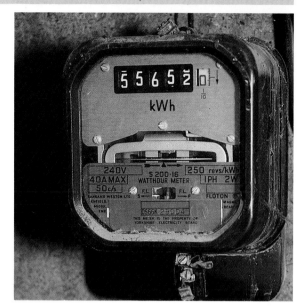

Fig. 14.35 An electricity meter

Yorkshire Electricity Board
PB XG2 Limewood Approach Leeds LS15 8TB
V.A.T. REGISTRATION NUMBER 238 5679 21

If you have a query on this account please ring

MR LAMBERT
ILKLEY

SUPPLY TO
MR A J LAMBERT
ILKLEY

Payment facilities —See overleaf   meter reading date 2 MAR 1983   tax point and date of issue 4 MAR 1983

Tariffs and meter readings ('E' - estimated, 'R' - customer's own reading)

present	previous	units supplied		fixed charges £	amount £	VAT rate %
53147	52362	785				
DOMESTIC GD			785 @ 5.10P	6.35	46.38	0

AMOUNT NOW DUE AND PAYABLE £ 46.38

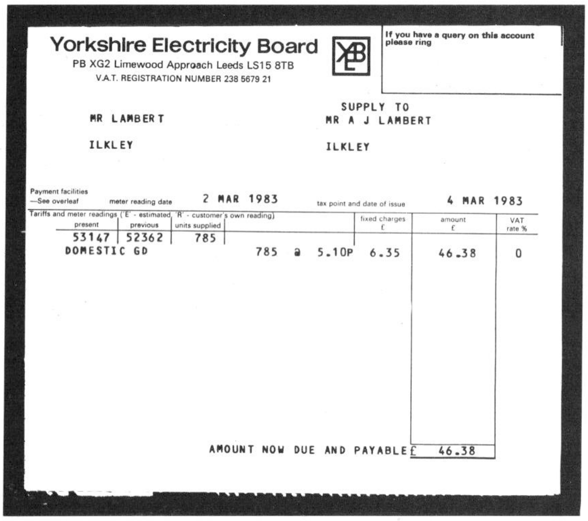

Fig. 14.36 An electricity bill

It is electrical appliances which are designed for heating things which are expensive to run. For example, if a 3 kW immersion heater is left on for 4 hours, it will convert 3 kW × 4 hours = 12 kW h of electrical energy to internal energy of the water. If electricity costs 5.1p per unit as in the bill illustrated, then this will cost 12×5.1p = 61.2p.

Compare this with a 60 W (0.06 kW) light bulb left on for the same time. It will convert 0.06 kW × 4 hours = 0.25 kW h of electrical energy. This will cost $0.24 \times 5.1 = 1.224$p. It is well worthwhile insulating hot water tanks and buildings so that heat energy is not wasted.

If it takes 3 minutes to boil the kettle of water in Fig. 14.27, how much would this have cost in March 1983? (Remember to convert 3 minutes to a fraction of an hour.) Find out how much a unit of electricity costs today. How much does it now cost to run an immersion heater for 4 hours, a 60 W lamp for 4 hours, and to boil the kettle?

14.12 Measuring the specific heating energy of aluminium

The idea of specific heating energy was first introduced in Section 9.11. You will remember that in order to measure the specific heating energy of anything you must provide a known quantity of energy to a known mass of the substance and measure the resulting temperature change, the formula connecting all these quantities being

$$\text{energy} = \text{mass} \times \text{specific heating energy} \times \text{temperature change}$$

One way of obtaining the necessary energy is to use an electric immersion heater. The electrical energy which is converted into internal energy can be measured accurately.

Fig. 14.37 shows a suitable apparatus, but there is at least one important way in which it could be improved; can you think what this is?

Assemble this apparatus (with improvements if you can think of them). The immersion heater will probably be designed to work off a 12 V supply,

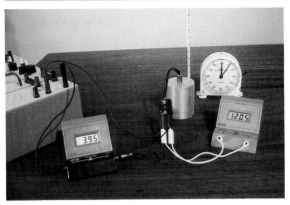

Table of results	
Potential difference across heater	V
Current through heater	A
Time for which experiment was run	s
Temperature at the start of timing	°C
Temperature at the end of timing	°C

Fig. 14.37　Measuring the specific heating energy of aluminium

but make sure you use the correct supply for your heater. Switch on the power to the immersion heater, and record the current through the heater and the potential difference across the heater in a table like the one above. Wait a minute or so until the heater itself has had time to warm up and the thermometer shows that the temperature of the aluminium is beginning to increase, then start the stopclock and at the same time note the temperature of the aluminium. After a convenient time (e.g. 100 s) note the new temperature of the aluminium.

Now answer these questions.

1 How do you calculate the *power* of the heater from these results? What is the power of your heater?

2 How do you go on to calculate the *total electrical energy supplied*? If all this energy becomes internal energy in the aluminium, how much energy was supplied by your heater?

3 What was the change in temperature of your piece of aluminium?

4 You cannot yet use the equation at the start of this section to calculate the specific heating energy because there is one more thing to measure. What is it?

5 Make this other measurement and then use the equation at the start of this section to calculate the specific heating energy of aluminium.

SUMMARY

Now that you have finished studying this chapter on electrical energy, there are a number of things you should know and be able to do.

1 You should:

a) know the definition of a coulomb of electric charge;

b) know the definition of a potential difference of 1 volt, and be able to use this definition in simple calculations involving electrical energy;

c) know that electric power can be calculated from the equation power = volts × amps, and be able to use this equation in simple calculations;

d) be able to use an ammeter correctly;

e) be able to use a voltmeter correctly;

f) understand how electricity is distributed in a house;

g) know how a fuse is used to protect equipment, and be able to choose the correct fuse for a particular task;

h) understand the reason for earthing mains equipment;

i) be able to perform simple calculations involving the cost of electricity;

j) be able to describe one experiment to measure specific heating energy using electrical energy.

2 You should know what each of the following is, or what each does:

analogue meter	digital meter
analogy	ammeter
voltmeter	potential difference
distribution box	fuse box
ring main	circuit breaker
kilowatt hour	

3 These are some of the other words that have been used in this chapter. You should know what each word means:

frayed	traditional
appliance	insulate

FURTHER QUESTIONS

1 A pupil set up the circuit illustrated in Fig. 14.38 in order to measure the power of a lamp.

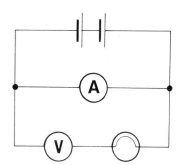

Fig. 14.38

a) Write down everything that is wrong with this circuit.
b) If this were to be connected, what would happen to i) the cells, ii) the lamp?

2 Here are three household appliances:
   a television rated at 240 W;
   an electric fire rated at 1 kW;
   an electric iron rated at 720 W.
a) What current passes through each appliance when connected to a 240 V mains supply?
b) What is the rating (3 A or 13 A) of the fuse that should be used in the mains plug connected to each appliance?
c) Explain whether it would be safe to run all three appliances together from one 13 A socket using a three-way adaptor.

3 If electricity costs 6p per kW h, how much does it cost to leave a 3 kW fire and five 100 W light bulbs on for 10 hours during the night?

4 A householder uses an immersion heater rated at 3 kW, eight lamps each rated at 100 W and a TV set rated at 200 W for a total time of 8 hours; this costs the householder £1.60. What would have been the cost if only the lamps and the TV had been used?

5 An electric immersion heater increases the temperature of 500 g of water from 20 °C to 50 °C in 5 minutes. The specific heating energy of water is 4200 J/kg °C.
a) Assuming there are no energy losses, what is the power of the heater?
b) If the heater is working from a 240 V supply, what is the current passing through the heater?

6 An electric motor raises a load of 5 kg a distance of 2 m in 5 s. While doing this, the p.d. across the the motor is 20 V and the current through the motor is 3 A.
a) What is the electric power input to the motor?
b) How much gravitational potential energy does the load gain?
c) What is the output power of the motor?
d) What is the efficiency of the motor?
e) Suggest possible reasons for the efficiency being less than 100%.

7 A saucepan on an electric hotplate contains 1 kg of water at 20 °C. The hotplate is switched on and it takes 5 minutes to bring the water to the boil. The power of the hotplate is 2 kW.
a) How much energy was transferred by the hotplate in 5 minutes?
b) How much energy was required to bring 1 kg of water to the boil from 20 °C?
c) Explain why the answers to a) and b) are different.

8 Fig. 14.39 shows three lamps lit, each to normal brightness, by connecting them in series to a 240 V mains supply. The ammeter reads 0.5 A.

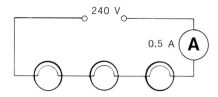

Fig. 14.39

a) What is the p.d. across each lamp?
b) What is the power output from each lamp?
c) For how long can this circuit be switched on before consuming 1 kW h of energy?
d) What will happen to the other lamps in this circuit if:

   i) the filament in one lamp breaks?
  ii) a short circuit occurs in one of the lamps?
e) Explain how, if at all, a suitable fuse can be of use in either of the situations in part d).

9 A householder decides to go to bed by torch light in order to save electricity. Assume that the torch contains two 1.5 V cells each of which costs 20p and that the average current through the lamps is 0.25 A. The cells last for 2 hours.
a) What is the power of a lamp?
b) What is the total energy provided by the cells during their life?
c) How many joules do the cells provide for one penny?
d) How many joules are there in a kilowatt hour?
e) If the cost of 1 kilowatt hour provided by the electricity board is 5p, how many joules are obtained for 1p?
f) Was this householder's idea a good one?

15

ELECTRICAL RESISTANCE

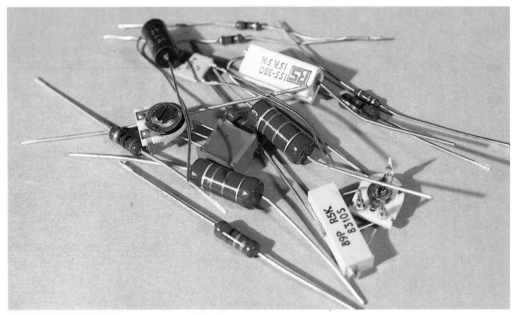

Fig. 15.1   Resistors come in many shapes and sizes

15.1   Introduction

You have already met the idea of 'resistance' in this book. At the start of the last chapter you were reminded that the more lamps there are in a series circuit, the more 'difficult' it is for the electric current to pass round the circuit. In other words, the more lamps there are, the greater the 'resistance' of the circuit.

The photograph shows a lot of different resistors. Why should anybody want these resistors? Since electrical energy will be converted (To what form?) as current passes through them, surely energy is going to be wasted if these resistors are deliberately put into a circuit? What is the point of them? Do they all behave in the same way? They certainly do not all *look* the same.

In this chapter you will look at the way in which physicists measure resistance, how resistors are sometimes used, and some of the factors which affect the resistance of a substance.

15.2   You should know . . .

In order to understand this chapter, you should know and understand:
1 how to set up simple electric circuits by following circuit diagrams;
2 how to use an ammeter and voltmeter correctly;
3 that the resistance of anything depends on its shape—a long thin wire has more resistance than a short fat wire;

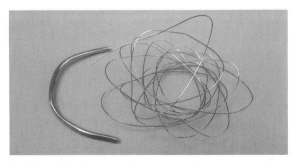

Fig. 15.2   Which wire has the greater resistance?

4 that the resistance of anything depends on what it is made of (a piece of carbon has more resistance than a piece of copper of the same size and a piece of glass of the same size has much more resistance than either copper or carbon);

5 that a substance with a high resistance is called an insulator.

15.3 Ohm's law

The object of this section is to find a way of measuring resistance. To do this, you will have to look at how the *current* flowing through a conductor varies with the *potential difference* across that conductor.

You may have some suitable ready-made resistors available, but if not, use about a metre of 28 swg eureka wire as a resistor. Coil it round a pencil to make it neat, and hold it in a clip component holder as shown in Fig. 15.3, which also shows the rest of the apparatus needed.

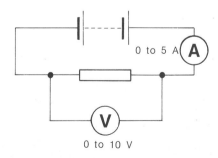

Fig. 15.3   How to investigate the relationship between the potential difference across a wire and the current through the wire

Fig. 15.4   The circuit diagram for Fig. 15.3

Drive the current round your circuit using one, then two, then three cells, and so on up to about six cells. On each occasion, record the current flowing and the p.d. across the wire in a table like the one below. (Alternatively, you can use six different voltages from a variable voltage power supply instead of using cells.) Copy out the table below and write your results in your table.

Take care not to let the wire become hot. Disconnect the circuit after every reading.

p.d. across wire (V)			
Current through wire (A)			

As usual, it is sensible to draw a graph of the results. (Why is it sensible?) Plot p.d. up the *y*-axis and current along the *x*-axis as in Fig. 15.5, which shows a graph of the results Kate obtained when she did this experiment. Does your graph look similar to this one? In particular, is your graph a *straight line* passing through the *origin*? If so, you can come to the same conclusion as Kate, that the current through your wire is *proportional* to the p.d. across the wire.

It turns out that this relationship between current and p.d. is true for all metals, and for carbon. It is called **Ohm's law**.

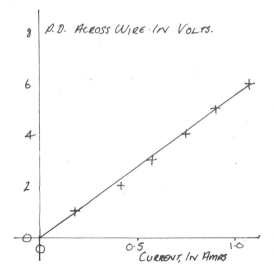

Fig. 15.5   Kate's graph of p.d. against current

The ratio p.d./current is called the **resistance** of the wire. Look again at Fig. 15.5. You can see that a p.d. of 5.5 V drove a current of 1 A through the wire, so the resistance of the wire was

$$\frac{5.5\,\text{V}}{1\,\text{A}} = 5.5 \text{ units of resistance}$$

The unit of resistance is the **ohm**, (symbol Ω). The resistance of this wire was 5.5 ohms.

Since the graph is a straight line passing through the origin, the ratio p.d./current (which is the gradient of this graph) is the same no matter where on the graph you measure it; in other words, the resistance of the wire is constant, it does not depend on the p.d. across the wire (in contrast to some other things you will meet later).

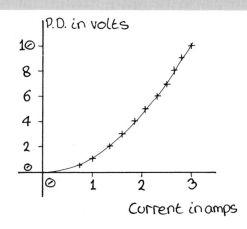

Fig. 15.7   Toby's graph of p.d. against current for a bulb. How does this compare with Kate's graph in Fig. 15.5?

15.4   The effect of temperature on resistance

In the last experiment you were instructed not to let the wire become hot. Now find out what happens if a wire *does* become hot. Repeat the experiment using a 12 V car light bulb, applying potential differences from 0 V to 12 V in steps of 0.5 V so that the filament gradually becomes hotter. Draw a graph of p.d. against current in the same way as in the first experiment.

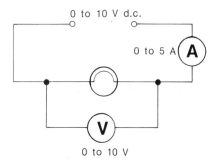

Fig. 15.6   How to investigate the resistance of a bulb

Toby's results for such an experiment are shown in Fig. 15.7. Do you obtain a graph with the same shape? Does this graph show that the resistance of the filament becomes greater as it becomes hotter, or less as it becomes hotter?

What current was flowing through Toby's filament when a p.d. of 0.5 V was applied across it? What, then, was the resistance of the filament? What was the resistance of the filament when a p.d. of 12 V was applied across it?

All metals behave in the same way as the filament in Toby's (and your) lamp. The *complete* version of Ohm's law is therefore:

The current through a conductor is proportional to the potential difference across the conductor, provided the temperature remains constant.

Although all metals obey Ohm's law, do not imagine that everything obeys this rule. The substances that do *not* obey Ohm's law are of great importance to us today, for example, in electronics, as you will see soon.

15.5   Switching on a lamp

Have you ever noticed that when a light bulb 'blows' it usually happens when you switch it on, rather than when it has been on for some time? Why is this? You have just learned that when the filament of a lamp is cold, its resistance is low. As it becomes hotter the resistance increases. Perhaps as you switch on the lamp there is a comparatively large current flowing. After a very short time, this might decrease to a lower value as the resistance of the filament increases. If this is so, there will be a surge of current through the lamp as you switch it on, and this could be responsible for 'blowing' the light bulb.

How can we investigate if there is a surge of current (and hence a surge of power) when you first switch on a lamp. Any surge will be finished in a very short time (less than half a second) so we need an instrument which will measure the current in a circuit many times in the first 0.5 second or so, and 'remember' (or store) those values of current so that we can read them later.

A VELA is programmed to take many measurements very rapidly, and Fig. 15.8 shows this instrument being used to measure the current in a lamp every millisecond for the first second after the current is switched on. The diagram of the circuit is shown in Fig. 15.9.

Fig. 15.8 This apparatus is being used to record the surge of current which occurs when a bulb is switched on

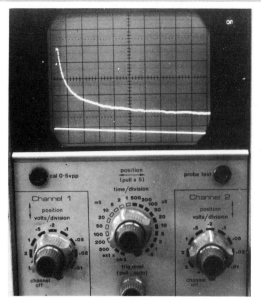

Fig. 15.10 The results of the experiment illustrated in Fig. 15.8 displayed on an oscilloscope

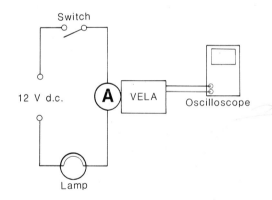

Fig. 15.9 The circuit diagram of the apparatus in Fig. 15.8.

After the VELA had finished recording, the readings were output repetitively to an oscilloscope. The trace on the oscilloscope screen is shown in Fig. 15.10; it is showing a graph of how the current through the circuit was varying with time. You can see that there was a large surge of current at the start, when the circuit was first switched on, and that the current rapidly decreased to a lower, constant, value. The oscilloscope shows how the current varied in the first second. Roughly how long was it before the current had settled down to a constant value?

15.6 Resistors in series and parallel

If several resistors are connected in series, the combined resistance must be *higher* than that of any one of the resistors by itself. (Why?) In Fig. 15.11, the resistors of 4 ohms, 6 ohms and 10 ohms have combined resistance of 20 ohms, and exactly the same current would flow if they were replaced by a resistor of value 20 ohms.

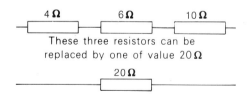

Fig. 15.11 Combining resistors in series

In general, a number of resistors of value R_1, R_2, R_3, etc. in series hve a combined resistance of R given by

$$R = R_1 + R_2 + R_3 + \text{etc.}$$

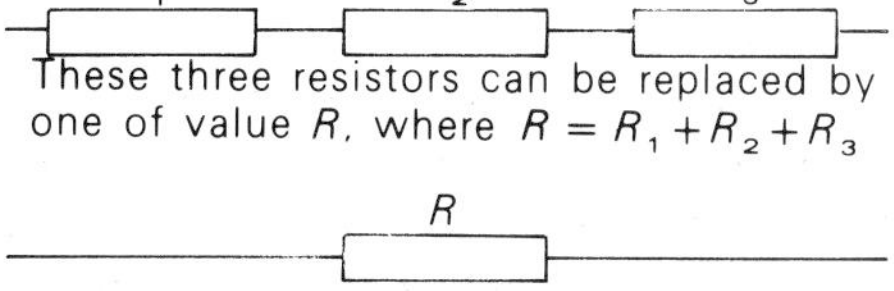

Fig. 15.12 Combining resistors in series

173

It is not quite so obvious what the combined effect of resistors in parallel will be. Look at Fig. 15.13.

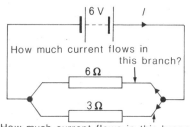

Fig. 15.13   Resistors in parallel

The two resistors provide *alternative paths* for the current, so it must be 'easier' for the current to pass round this circuit than it would be with only one of these resistors in the circuit. The value of the combined resistance must be less than that of either of the resistors, i.e. less than 3 ohms in this case.

Suppose these resistors were joined to a 6 V battery, as shown in the diagram.

1 What current flows through the 6 ohm resistor? (Use $V = IR$.)
2 What current flows through the 3 ohm resistor?
3 What is the total current that flows from the battery round the circuit?
4 What *single* resistor would you put in the circuit in place of the two resistors in parallel to cause the same current to flow?

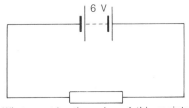

What must be the value of this resistor so that the same current flows in this circuit as in the circuit in Fig. 15.13?

Fig. 15.14   What is the value of the single resistor which replaces the two parallel resistors?

5 Is the value of this single resistor less than 3 ohms, as we suggested it ought to be?

We can extend this argument to any number of resistors in parallel. Fig. 15.15 shows three resistors R_1, R_2, and R_3 in parallel. The total current leaving the battery and flowing round the circuit is I; at the junction it splits so that I_1 goes through resistor R_1, and so on as in the diagram. The potential difference

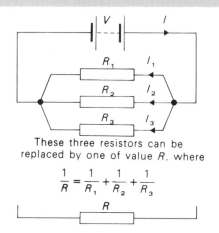

Fig. 15.15   Several parallel resistors

across all three resistors is V. (Why is it the same for all three?)

Rearranging the equation $V = IR$ to give $I = V/R$, and applying this to each of the three resistors in turn, gives the following three equations:

$$I_1 = \frac{V}{R_1}$$

$$I_2 = \frac{V}{R_2}$$

$$I_3 = \frac{V}{R_3}$$

But $I_1 + I_2 + I_3 = I$ (the total current). (Why?)

Therefore:

$$I = \frac{V}{R_1} + \frac{V}{R_2} + \frac{V}{R_3}$$

If the three resistors were replaced by one resistor of the same resistance R as the three combined, all the current I would pass through this resistor, and

$$I = \frac{V}{R}$$

Therefore:

$$\frac{V}{R} = \frac{V}{R_1} + \frac{V}{R_2} + \frac{V}{R_3}$$

Dividing through by V gives

$$\frac{1}{R} = \frac{1}{R_1} + \frac{1}{R_2} + \frac{1}{R_3}$$

This can be extended to any number of resistors in parallel.

Thus, if the three resistors illustrated in Fig. 15.11 were to be combined in parallel instead of in series, their combined resistance R would be given by

$$\frac{1}{R} = \frac{1}{4} + \frac{1}{6} + \frac{1}{10}$$

$$= \frac{15 + 10 + 6}{60}$$

$$= \frac{31}{60}$$

Therefore:

$$R = \frac{60}{31}\,\Omega$$

$$= 1.94\,\Omega$$

(As always, a lower value than any of the individual resistors.)

15.7 The potential divider

The wire in Fig. 15.16 has been connected across a 5 V power supply. As you learned in the last chapter, each coulomb of electric charge which comes from the power supply carries 5 J of electrical energy. Each coulomb returning to the power supply carries no energy. The energy has gone to increasing the temperature of the wire.

The right end of the wire has been labelled '5 V' to remind you of the energy of each coulomb at that point. If the wire is a uniform one, halfway along, half the electrical energy will have been converted, and each coulomb will only have 2.5 J of electrical energy. The halfway point has therefore been labelled '2.5 V'.

If you need electricity at a potential of 2.5 V for any purpose, you can 'tap off' the electricity at this point as shown in Fig. 15.17.

[This will only apply accurately if the current you tap off is negligible, so that the current i in both halves of the wire is the same. Then the potential difference ($= iR$) across both halves of the wire will be the same. See Fig. 15.18.]

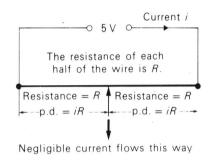

Fig. 15.18

Where, along this wire, would you tap off electricity at 1 V, and at 3 V?

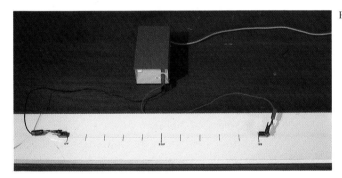

Fig. 15.16   A simple 'potential divider'

Fig. 15.17   Tapping off 2.5 V from the potential divider

When the wire is being used in this way we say it is a **potential divider**. Can you see why it is called this? In practice, you would use a **potentiometer**, as illustrated in Figs 15.19 and 15.20 to tap off whatever voltage was required. Alternatively, if the voltage output of the potential divider circuit did not need to be altered, then two fixed value resistors, as illustrated in Fig. 15.21, could be used.

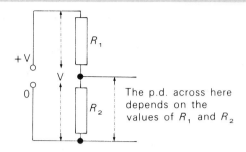

Fig. 15.21   A potential divider formed of fixed resistors

Fig. 15.19(a)   A potentiometer

Fig. 15.19(b)   Inside the potentiometer

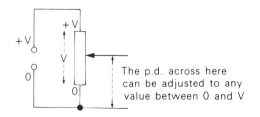

Fig. 15.20   The symbol for a potentiometer

15.8   Shunts and multipliers

When you use ammeters or voltmeters, you may have used a pattern similar to that illustrated in Fig. 15.22, where you have to plug in the correct **shunt** (for an ammeter) or **multiplier** (for a voltmeter). What is in a shunt or multiplier that enables the *same* meter to read either current or potential difference, and to have such a wide range of sensitivities?

Fig. 15.22   A typical moving-coil meter

Let us use the meter illustrated in Fig. 15.22 as an example. As shown, the meter has a maximum reading of $100\,\mu A$, or a 'full scale deflection' of $100\,\mu A$, to use the usual phrase. (What fraction of 1 A is $100\,\mu A$?) It has a resistance of $1000\,\Omega$, so the potential difference across the meter when a current of $100\,\mu A$ (0.0001 A) is flowing through it is given by

$$V = IR$$
$$= 0.0001\,A \times 1000\,\Omega$$
$$= 0.1\,V$$

So as it stands, the meter could also be used as a voltmeter reading up to 0.1 V.

Suppose we want to convert this meter to an ammeter with full scale deflection of 1 A. We clearly cannot send 1 A through the meter as it stands. (What do you think will happen to the meter?) Most of the current must be 'shunted round' the meter, as shown in Fig. 15.23, and only a small proportion must be allowed through the meter. If 1 A is flowing towards the meter, $100\,\mu A$ (0.0001 A) can go through the meter (which will move to full scale deflection, which we will then read as '1 A') and the other 0.9999 A must be shunted round the meter, as shown in the diagram.

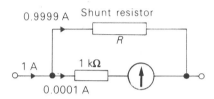

Fig. 15.23   Shunting current round a meter

Since the p.d. across the meter is 0.1 V, the p.d. cross the shunt resistor will also be 0.1 V. (Why?) Using $V = IR$ to find the value of the shunt resistor:

$$0.1\,V = 0.9999\,A \times R$$

$$R = \frac{0.1\,V}{0.9999\,A}$$

$$= 0.10001\,\Omega$$

This is a very small resistance. It is often made from an etched strip of copper. Fig. 15.24 shows the inside of a 100 mA shunt designed to go with the meter in the previous photograph.

Fig. 15.24   Inside a 100 mA shunt for the meter in Fig. 15.22. The copper tracks form the shunt resistor

The total resistance of the ammeter is $0.1\,\Omega$ ($0.10001\,\Omega$ and $1000\,\Omega$ in parallel). This is likely to be lower than any other resistance in the circuit, so including an ammeter like this in a circuit will not significantly alter the circuit. This is an important point in *any* scientific measurement; your measuring instrument must not disturb or alter the system you are measuring. Good ammeters have a very low resistance compared to the rest of the circuit.

Suppose you wanted to use the same meter as a voltmeter reading up to 10 V. Clearly you cannot put 10 V across the meter as it stands. How much current would flow through the meter if you did do such a foolish thing?

Only a small fraction of the 10 V must be across the meter, the rest must be across a resistor in series with the meter, as shown in Fig. 15.25. This series resistor must have 9.9 V across it. (Why?) The current through both the resistor and the meter is 0.0001 A when the meter is at full scale deflection, so we can now use $V = IR$ to find the resistance of this series resistor:

$$9.9\,V = 0.0001\,A \times R$$

$$R = \frac{9.9\,V}{0.0001\,A}$$

$$= 99\,000\,\Omega \text{ or } 99\,k\Omega$$

This, of course, is a very large resistance. Together with the $1\,k\Omega$ resistance of the meter itself, the total resistance of the voltmeter is $100\,k\Omega$. Good voltmeters have a large resistance, so that the extra current flowing round the circuit (0.0001 A in this case) as a result of placing the voltmeter in the circuit is very small. Modern digital voltmeters usually have resistances greater than a megohm. (How many ohms is that?)

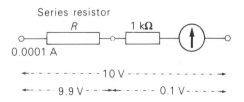

Fig. 15.25   Converting a basic meter to a voltmeter, using a series resistor

15.9 The internal resistance of a cell

Connect a single cell (preferably a fairly old one) to a lamp and measure the voltage across the cell [Fig. 15.26(a)]. Make a note of this voltage. Now connect a second lamp to the cell, parallel to the first, and note the voltage again. Repeat this with a third lamp.

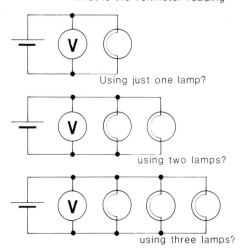

Fig. 15.26 Will the voltmeters in these three circuits read the same, or different?

Why does the voltage go down a little each time you add another lamp? Since you are adding the lamps in parallel, you are drawing more current from the cell each time you add a lamp (check this using an ammeter if you wish). Not only does this current have to go through the lamps, it also has to go through the cell. It requires a little energy to push the current through the cell; the more the current, the more the energy required. Some of the electrical energy given to the electric charge by the cell is 'used up' in passing through the cell itself. In other words, the cell itself has a resistance; we call this the **internal resistance** of the cell.

This internal resistance effectively limits the maximum current you can draw from the cell. Suppose, for example, the internal resistance of a 1.5 V cell was 0.5 ohm. Even if the terminals of the cell were short circuited with a wire of negligible resistance, the current flowing would be given by

$$i = \frac{V}{R}$$
$$= \frac{1.5\,V}{0.5}$$
$$= 3\,A$$

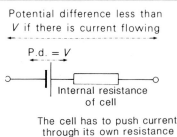

Fig. 15.27 The internal resistance of a cell

A single U2 cell might have an internal resistance of about 0.5 ohm. The internal resistance of a cell increases as it becomes discharged.

Note that rechargeable batteries have a low internal resistance, and care should be taken not to 'short' their terminals. What do you think would happen if the terminals were shorted?

All sources of potential difference have an internal resistance. It ranges from a fraction of an ohm for a typical transformer used in a laboratory low voltage supply to hundreds of ohms for a solar cell.

15.10 Some special resistors

The little device shown in Fig. 15.28 is a resistor with a very special property. It is not made from a metal or carbon, like the resistors you have used until now, but it is made from a compound called cadmium sulphide.

Fig. 15.28 What sort of resistor is this?

Include this device in the simple circuit shown in Fig. 15.29. Make the appropriate measurements of p.d. and current and calculate the resistance of this device. Now cover it up with your hand, so that little light falls on it. What happens to the current in your circuit? What has happened to the resistance of the device? What happens to its resistance if you shine a bright light, or sunlight, on it?

Fig. 15.29   How to investigate a light-dependent resistor in an electric circuit

This device is usually called a **light-dependent resistor** (or LDR for short). It should be obvious why! Can you think of any practical use for a light-dependent resistor?

Fig. 15.30 shows another interesting kind of resistor; it is called a **thermistor**. Like most other

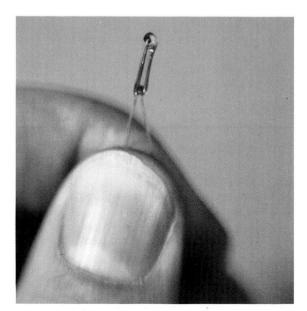

Fig. 15.30   A thermistor

resistors, its resistance depends on its temperature, but not in the same way as the resistors you have used before. Connect a thermistor into a circuit like the one illustrated in Fig. 15.31 and measure its resistance. Then put the thermistor into warm water for a minute in order to increase its temperature, taking care not to allow the connecting leads to enter the water (since water will conduct electricity). What happens to the current in the circuit? What is the value of the resistance now? In what way is this behaving differently from the metal resistors (such as the lamp filament) you used earlier?

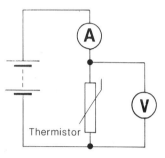

Fig. 15.31   A circuit for investigating the behaviour of a thermistor

Can you think of any sort of measuring instrument that could be based on a thermistor?

Both LDRs and thermistors are often used in potential divider circuits which you learned about in Section 7 of this chapter. Fig. 15.32 shows the LDR in a potential divider circuit, with a voltmeter to measure the 'output' voltage, that is, the voltage at the junction between the LDR and the fixed resistor.

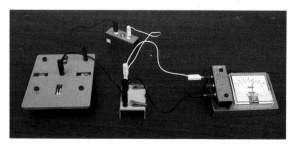

Fig. 15.32   A light-dependent resistor in a potential divider circuit

Assemble this circuit for yourself. Note the reading on the voltmeter. Cover up the LDR so that it is dark. What happens to the voltmeter reading? Explain why this happens.

In the next chapter you will learn how to use this simple circuit to switch on a light or an alarm bell, for example, when it becomes dark.

SUMMARY

Now that you have finished studying this chapter on electrical resistance, there are a number of things you should know and be able to do.

1 You should:
   a) know Ohm's law;
   b) know the definition of 1 ohm of resistance, and be able to use this definition in simple calculations involving electrical resistance;
   c) know the effect that temperature has on the resistance of a substance;
   d) be able to calculate the effective resistance of several resistors in series;
   e) be able to calculate the effective resistance of several resistors in parallel;
   f) understand what a potential divider is, and how it is used;
   g) be able to calculate resistance values for shunts and multipliers for meters;
   h) understand why a good ammeter has a low resistance, and a good voltmeter a high resistance;

2 You should know what each of the following is, or what each does:

insulator	conductor
internal resistance	light-dependent
thermistor	resistor

3 These are some of the other words that have been used in this chapter. You should know what each word means:

contrast	surge
succession	individual
etch	rechargeable

FURTHER QUESTIONS

1 a) Why are the connecting wires that you use when making electric circuits, made of copper rather than any other metal?
   b) The overhead electricity cables that are used for transmitting electricity across the country are usually made of aluminium, with a steel core. (The steel core is to strengthen the cable.) Why do you think aluminium is used, rather than copper?

2 In the simple circuit illustrated in Fig. 15.33:
   a) what is the current flowing in the circuit?
   b) what will be the readings on each of the three voltmeters?

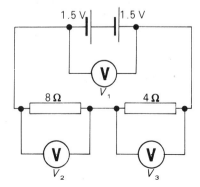

Fig. 15.33

3 An 8 V battery of negligible internal resistance is connected to resistors of 8 Ω and 16 Ω.
   a) What current flows if the resistors are:
      i) in series;
      ii) in parallel?
   b) Why do you need to know that the battery has 'negligible internal resistance'?

4 A 12 V car battery has an internal resistance of 0.05 Ω. It is used to turn a car's starter motor, which requires a current of 100 A.
   a) What p.d. is required to send a current of 100 A through a resistance of 0.05 Ω?
   b) What will be the p.d. across the battery when supplying a current of 100 A to the starter motor?
   c) Explain why any lamps which are switched on will become dim while the starter motor is turning.
   d) Suppose that as the battery becomes older, the internal resistance increases to 0.5 Ω. Explain what will happen when attempting to start the car.

5 In Section 15.8 you read that a measuring instrument, such as a voltmeter, should not disturb or alter the system on which a measurement is being taken.

a) Write down any examples of which you can think (excluding ammeters and voltmeters) where you have had to take care not to allow the measuring instrument to alter what was being measured.

b) Write down any examples of which you can think where the measuring instrument *did* 'disturb the system'.

6 A moving-coil ammeter has a resistance of $10\,\Omega$ and gives a full scale deflection when a current of $10\,mA$ passes through it.

a) How would you convert it to an ammeter with a full scale deflection of 1 A?

b) How would you convert it to a voltmeter with a full scale deflection of 10 V?

c) Suppose, in converting this meter to an ammeter with a full scale deflection of 1 A, you mistakenly plug in a shunt designed for a meter with a resistance of $1000\,\Omega$ and a full scale deflection of $100\,\mu A$, such as was described in Section 15.8.

  i) What would the true current be when the meter apparently read 1 A?

  ii) Do you think you can read the scale of the moving-coil ammeter in Fig. 15.22 with sufficient certainty for your mistake to matter?

7 You have to run a 12 V, 36 W lamp from an 18 V battery. You have available a variable resistor, ammeter, voltmeter as well as the lamp and battery.

a) Draw a circuit diagram showing how you would connect all the items together. Include the voltmeter to check the p.d. across the lamp, and the ammeter to measure the current in the circuit.

b) Is there an alternative place in the circuit you have drawn for:

  i) the ammeter;

  ii) the voltmeter?

If so, where?

c) What would you expect to be the reading on the ammeter when the p.d. across the lamp is 12 V?

d) What will be the value of the resistance of the variable resistor when the lamp is properly lit, i.e. with a p.d. of 12 V across it?

(You need to have studied Chapter 14 on electrical energy before you can attempt the following question.)

8 The circuit diagrams X, Y and Z in Fig. 15.34 show an electric heater connected in three different ways to a 240 V a.c. supply. The heater has two similar heating elements, A and B, each with a resistance of 60 ohms. There are three settings of the heater: low, medium and high.

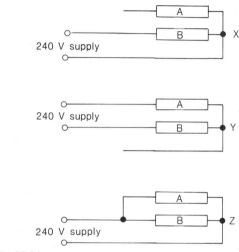

Fig. 15.34

a) Calculate the current drawn from the supply in each of the three circuits X, Y and Z.

b) Which circuit corresponds to the *high* setting and which corresponds to the *low*?

c) Calculate the power in circuit X.

d) If electrical energy costs 6p per kWh, calculate the cost of using the heater when connected as in X for 30 minutes.

The heater with its three settings is connected to the supply by a 3-pin plug which has a fuse inside it.

e) i) What is the purpose of the fuse?

  ii) How does it work?

f) Fuses marked 3 A, 5 A, 10 A and 13 A are available. Which would be the most appropriate one to use. Explain your choice.

g) Element A burns out. Discuss whether the heater would operate on each of the settings.

h) If in circuit A the voltage supply were halved, what effect would this have on the power?

(O&C Nuffield O-Level Physics, June 1983)

16

SEMICONDUCTORS

16.1 Introduction

What have the transistor, the integrated circuit and sand on the seashore in common? Answer: they all contain the element silicon. Silicon conducts electricity, although not nearly as well as any metal. It is called a **semiconductor**. (Germanium is another semiconductor.) In 1948 a method was invented of using a semiconductor to make the first transistor. Since then, advances in technology have enabled us to make circuits consisting of thousands of transistors, together with resistors and the connections between them, on a single small piece of silicon (Fig. 16.2) and hence has opened up the whole field of **microelectronics**. This is surely the fastest growing technology of the last quarter of the twentieth century.

In this book you have already seen some instruments which use microelectronics. Write down a few examples of such instruments. It would need many books to cover the whole field of electronics, so this chapter will concentrate on a few aspects of electronics which could be useful to us as physicists.

Fig. 16.1   What do the transistor, the integrated circuit and sand have in common?

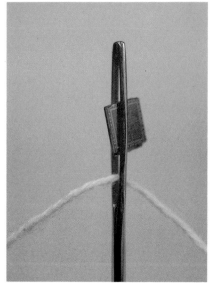

Fig. 16.2   A silicon chip. Many thousands of transistors and resistors are packed onto this tiny piece of silicon

16.2 You should know . . .

In order to understand this chapter, you should have read and understood Chapters 14 and 15 of this book. In particular, you should:
1 be able to connect electric circuits from diagrams;
2 be able to use an ammeter and a voltmeter;
3 understand how to measure resistance;
4 understand potential dividers.

16.3 The diode

Let us start by investigating the electronic component shown in Fig. 16.3. Inside the plastic casing is a semiconductor device made from silicon, called a **diode**. Will it conduct electricity well? Does it obey Ohm's law?

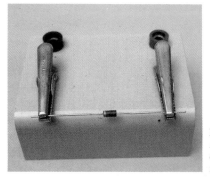

Fig. 16.3
A diode. This one is also made of silicon

There are many different types of diode. The one illustrated can safely carry a current of 1 A without overheating. Make sure you use a diode that can carry at least 1 A.

Fig. 16.4 shows Richard investigating the 'characteristics' of the diode. He is applying various potential differences across the diode and measuring any current that flows through the diode. Fig. 16.5 shows a diagram of the circuit he is using. The variable resistor is being used as a 'potential divider' in the way you learned in Chapter 15, so that very small p.d.'s can be applied to the diode.

Fig. 16.4   Richard investigating how the current through the diode depends on the p.d. across it.

Carry out this experiment for yourself. Does your diode behave in the same way, or does your diode have different 'characteristics'? Take care not to allow your current to exceed the maximum for your diode, or else you will ruin it.

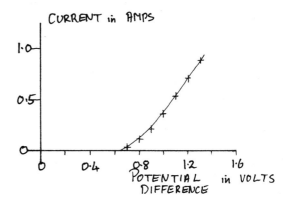

Fig. 16.6   Richard's graph of current against p.d. for a diode

The symbol for a diode is shown in Fig. 16.7, and the next illustration shows the same circuit as Fig. 16.5 using the correct symbol for the diode; it is connected so that it conducts easily. The 'arrow head' is called the **anode** and the 'bar' called the **cathode**. In the diode illustrated in Fig. 16.3, the cathode is identified with a ring painted round that end.

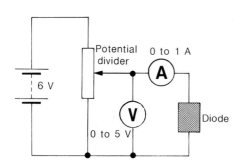

Fig. 16.5   The circuit that Richard is using in the previous photograph

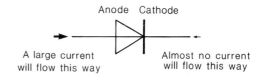

Fig. 16.7   The symbol for a diode

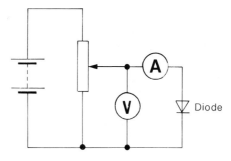

Fig. 16.8   Fig. 16.5 drawn using the correct symbol for a diode

He tried the diode both ways round in the circuit and found that it conducted much better in one direction than in the other. In fact, he did not think it conducted at all in one direction until he used a very sensitive ammeter and managed to detect the few microamps that were flowing. A graph of his results is shown in Fig. 16.6. Notice he found that, even with the diode connected so that it did conduct electricity easily, it did not start conducting until a p.d. of 0.7 V was applied.

To summarize: a diode has a *low* resistance in one direction, so it conducts electricity easily, provided the p.d. across the diode is greater than a certain value (about 0.7 V for a silicon diode). In the opposite direction it has a very high resistance and hardly any electricity will pass this way.

16.4 What use is a diode?

The current from the mains is alternating current (a.c.), that is, the current flows backwards and forwards many times a second. In Britain, the frequency of the mains is 50 Hz. However, many things, such as electronic instruments and small electric motors, need direct current (d.c.) at a low voltage. Changing the mains voltage of 240 V to a lower voltage requires a transformer, which you will learn about in Chapter 18. A diode can be used to change the alternating current to direct current.

Look at the simple circuit in Fig. 16.9. Although the circuit is being supplied with a.c., the diode only allows the current to flow in one direction, so direct current flows through the lamp. It is not very *smooth* d.c. Fig. 16.10 shows an oscilloscope connected across the lamp. You can see that for half the time there is no current flowing at all. Why does this happen.

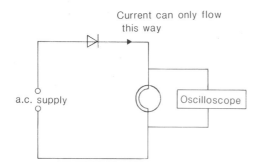

Fig. 16.9   A simple rectifier circuit

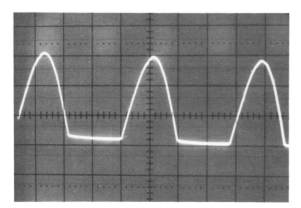

Fig. 16.10   A trace on an oscilloscope connected as shown in Fig. 16.9

Converting a.c. to d.c. in this way is called **rectification**. This particular circuit is a **half-wave rectifying** circuit. Why do you think it is called this?

Clearly a half-wave rectifying circuit is not very efficient. A better circuit is the **bridge rectifier**, which provides **full-wave rectification**. This is shown in Fig. 16.11. The output from this circuit is shown in the diagram. Try to work out for yourself how this arrangement of diodes gives the output shown.

The output from the circuit in Fig. 16.11 is not *smooth* d.c.; the value of the output voltage is clearly changing all the time. Look at Chapter 19 to find out how to smooth this d.c. using a capacitor.

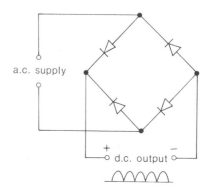

Fig. 16.11   A bridge rectifier circuit. How does it work?

16.5 Light-emitting diodes

Fig. 16.12   A tuning indicator on a radio made from light-emitting diodes

The row of lights at the top of this radio are diodes. They behave as any other diode, with a low forward resistance and a high reverse resistance. When current flows through these diodes, they give off light. The colour of the light depends on what has been added to the silicon during manufacture. For example, the red diodes are 'doped' with some gallium arsenide.

These special diodes are called **light-emitting diodes**, or LEDs for short. Do you think they have any advantage over ordinary filament lamps? If so, what advantage?

LEDs do not have to be circular like those in Fig. 16.12. A common arrangement is to put seven bar-shaped LEDs in one package as in the next photograph. By illuminating different combinations of bars, different numbers (and a few letters) can be formed. This is called a **seven-segment display**. Where else have you seen such a display in use, both inside and outside the laboratory?

Fig. 16.13 This seven-segment display is made from light-emitting diodes

Many LEDs need a current of about 10 mA to light up satisfactorily. Too much current will destroy the LED. Since a LED has a very low resistance when it is conducting, it must be used with a resistor in series to limit the current. For example, suppose you wish to light a LED from a 6 V battery. What value of resistor must you use in series with the LED to limit the current to 10 mA?

Look at Fig. 16.14. The p.d. across a LED when it is conducting 10 mA is about 1 V. (This is a result of the way a LED is made.) Therefore there must be 5 V across the resistor R. There will also be 10 mA flowing through R. (Why?)

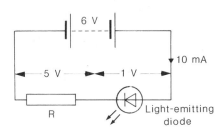

Fig. 16.14 How to calculate the value of a series resistor for a LED

Using
$$V = IR$$
$$R = \frac{V}{I}$$
$$= \frac{5\text{ V}}{0.01\text{ A}}$$
$$= 500\ \Omega$$

You must use a $500\ \Omega$ resistor in series with the LED.

16.6 Investigating a transistor

Fig. 16.15 Transistors come in many shapes and sizes

Transistors come in a vast range of sizes and types designed for different purposes. However, they all behave in broadly the same way. The majority consist of *three* layers of semiconductor, either 'n-type' sandwiched between two pieces of 'p-type' (a **pnp-transistor**) or 'p-type' sandwiched between two pieces of 'n-type' (an **npn-transistor**). (The meaning of the phrases 'n-type' and 'p-type' need not concern us here; they refer to the way in which the semiconductor has been prepared for use in a transistor.) The general construction of an npn-transistor is shown in Fig. 16.16. Notice the names of the various parts. The circuit symbol for each kind of transistor is in the next diagram.

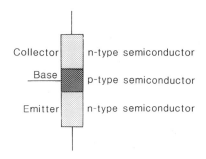

Fig. 16.16 A simplified diagram of an npn-transistor

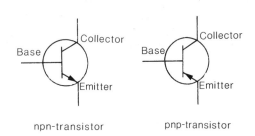

Fig. 16.17 The circuit symbol for transistors

Start an investigation into the way a transistor behaves by connecting the circuit shown in Fig. 16.18. Notice that, at this stage, the base terminal is not connected to anything; you are only using the collector and emitter.

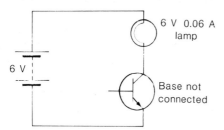

6 V 0.06 A
lamp

6 V

Base not
connected

Fig. 16.18   A circuit for starting to investigate the behaviour of
a transistor

How you assemble this circuit depends on what you have available in your lab. The easiest way is to have the circuit ready-assembled for you and mounted on a suitable board as in Fig. 16.19.

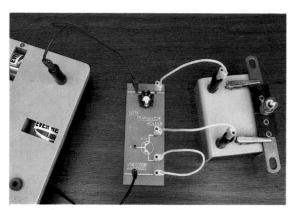

Fig. 16.19   A ready-made version of the circuit in the previous
diagram

An alternative is shown in the next photograph, where the circuit has been assembled on 'Proto-board'. This is much more fiddly than the first method, with plenty of opportunity for wrong connections! If you become sufficiently interested in electronics to want to go on to design and build other circuits, you may well find protoboard very useful. You will need to know which is the base, emitter and collector of your transistor if you use this sort of system, and take care to connect them correctly, or you will ruin your transistor.

When you switch on the circuit, does the lamp light up? Does any current flow in your circuit? (Include an ammeter as well as the lamp in your circuit if you wish.) Do you think the resistance of the transistor is high or low?

Fig. 16.20   The same circuit assembled on 'Protoboard'

Now add an extra part to the circuit, as in Figs 16.21 and 16.22, so that you can direct a small current into the base of the transistor. The $5\,k\Omega$ resistor is to ensure that the base current *is* small. The $0–1\,mA$ meter is needed to measure this small current. The current for this part of the circuit could come from another battery, but it is much more convenient to use a potential divider across the original battery as shown. This allows different voltages, and hence different currents, to be applied easily.

Adjust the potential divider so that *no* current flows onto the base of the transistor, then gradually increase the current. What current has to flow onto the base of the transistor for there to be sufficient current through the lamp to enable it to light up?

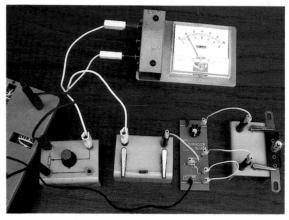

Fig. 16.21   Making a connection to the base of the transistor

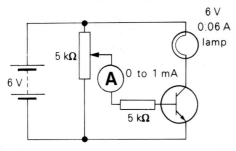

Fig. 16.22   A connection has been made to the base of the
transistor

Is the base current small or large compared to the
current through the lamp (the collector current)?

Copy out and complete the following sentence:

'A _____ current flowing onto the base of a
transistor can control a much _____ current
flowing onto the collector.'

Another way of describing the action of a transistor is to say that it is switched *off* when the
voltage applied to the base of the transistor is low
(and hence the base current is low), and it is
switched *on* when the voltage applied to the base is
high. Do you think this is a good description of the
way your circuit behaves?

16.7   The transistor as a switch

In this section there are a few practical circuits you
can try building which make use of the switching
action of a transistor. These circuits also use the
potential divider that you learned about in the last
chapter.

Look at Fig. 16.23. When light is falling on the
light-dependent resistor, will its resistance be high
or low? Will the voltage át point X therefore be high
or low?

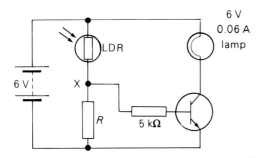

Fig. 16.23   How to use a light-dependent resistor and
transistor in a simple switching circuit

This voltage is connected to the base of the
transistor. When light is falling on the LDR, will the
transistor be switched *on* or *off*? Will the lamp be on
or off?

Now explain what will happen if the LDR is in the
dark.

The resistor *R* is usually a *variable* resistor. Why
do you think this is so?

Now assemble the circuit illustrated in Fig. 16.23
and see whether it behaves in the way you have
predicted in your answers to the questions above.
Can you think of a practical use of this circuit?

Fig. 16.24 shows almost the same circuit, but the
lamp has been replaced by a **relay**. What is a relay?

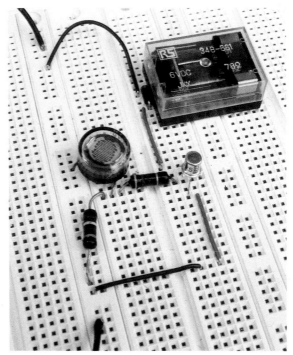

Fig. 16.24   The simple switching circuit can be used to operate
a relay

The relay could be used to switch on a more
powerful lamp, or perhaps switch on the circuit to a
warning bell.

Notice the use of a diode with the relay. When the
relay switches off, a high voltage can be generated
across the coil; the diode ensures the high voltage
does not reach the transistor.

Try to design and build a circuit which switches
on a warning bell when the temperature drops to
near freezing. (Use a thermistor—see Section
15.10.) How would you modify your circuit so that
a bell rings when a kettle boils?

16.8 The transistor as an amplifier

You have already learned that a small current flowing onto the base of a transistor results in a larger current flowing onto the collector. The transistor is acting as a **current amplifier**. How much amplification depends on the **gain** of the transistor. However, a single transistor does not make a very good amplifier, certainly not good enough for amplifying music, for example. A single transistor will distort the signal it is amplifying because the output will not be proportional to the input, and it will amplify low frequencies more than high frequencies. To make life even more difficult for the designer of an amplifier, the gain of individual transistors varies a lot, even with transistors of the same type.

A practical transistor amplifier will consist of many transistors and other components. In most cases, it does not make sense to build an amplifier using a lot of transistors. Integrated circuit amplifiers are very cheap and easy to use. Fig. 16.26 shows the internal design of the 'operational amplifier', as it is called, shown in the photograph. You can count the number of transistors for yourself! This 'op. amp.' costs less than 20p.

The amplifiers on your radio, hi-fi equipment, and laboratory equipment such as timer-scalers and VELAs are all likely to be op. amps. There is insufficient space in this book to give any details of experiments with op. amps. It would take a whole book to cover the subject in any reasonable depth! There are many electronics books available that will give you further details if you are interested.

Fig. 16.25 This operational amplifier contains the equivalent of over 20 transistors

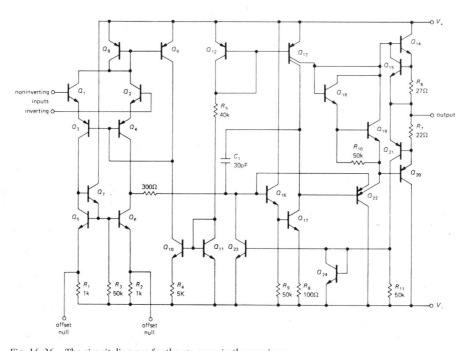

Fig. 16.26 The circuit diagram for the op. amp. in the previous photograph

SUMMARY

Now that you have finished studying this chapter on semiconductors, there are a number of things you should know and be able to do.

1 You should:
   a) know what a semiconductor is, and know one example of a semiconductor;
   b) know what is meant by microelectronics;
   c) know how a diode can be used to rectify alternating current;
   d) know the symbol for a diode;
   e) know the characteristics of a transistor;
   f) be able to use a transistor as a switch;
   g) know that a transistor can be used as an amplifier;
   h) be able to use a potential divider circuit to provide the input to the base of a transistor.

2 You should know what each of the following is, or what each does:

characteristics	cathode
anode	seven-segment display
relay	operational amplifier
bridge rectifier	half-wave rectifier
light-emitting diode	collector
base	emitter

3 These are some of the other words that have been used in this chapter. You should know what each word means:

component	instrument
concentrate	aspect
illuminate	sandwich

FURTHER QUESTIONS

1 Fig. 16.27 shows four different circuits containing various combinations of cells, lamps and diodes. Write down which lamps you would expect to be illuminated, giving reasons for your answers.

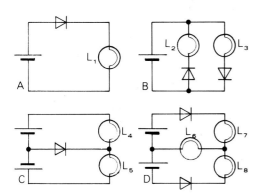

Fig. 16.27

2 Fig. 16.28 shows a simple moisture detecting circuit. The moisture detector consists of two metal rods placed close together.
   a) Describe how the warning lamp becomes illuminated when the moisture detector becomes wet.

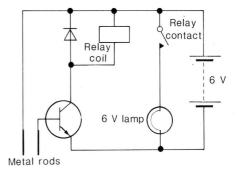

Fig. 16.28

b) The **current gain** of the transistor in the circuit in Fig. 16.28 is defined as

$$\frac{\text{current flowing onto the collector}}{\text{current flowing onto the base}}$$

If the relay requires a minimum current of 50 mA in order to work, and the current gain of this transistor is 100, what is the minimum current that must flow through the detector onto the base of the transistor?

c) How can you simplify this circuit so that the lamp will illuminate when the detector becomes wet without the use of a relay?

3 A 'puzzle box' (Fig. 16.29) contains two lamps and other simple components. When T_1 is connected to the positive terminal and T_2 to the negative terminal of a cell, lamp L_1 lights, but when the terminals are connected to the cell the other way round, lamp L_2 lights. Suggest what circuit is in the puzzle box.

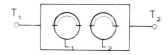

Fig. 16.29

4 Suggest how you could use a diode to protect a circuit against being accidentally connected to a cell incorrectly, that is, with the 'positive' lead connected to the negative terminal of the cell.

5 Design a simple circuit which uses a transistor, LDR and other components to ring an alarm bell when somebody walks through a door into a room. Are there any disadvantages of your simple circuit?

17

GENERATING ELECTRICITY

Fig. 17.1 How does the power station convert the energy stored in this coal to electrical energy?

17.1 Introduction

By now you should appreciate how important electrical energy is to you and the way you live. You know that a lot of our fossil fuels are used to generate electrical energy. But how does this power station convert the energy stored in the coal to electrical energy? You know part of what happens; you know about the energy changes which take place in order to drive the generator round. But how does a generator, whether a large one such as in the power station in Fig. 17.1, or the small one on the bicycle in Fig. 17.2, actually convert kinetic energy (of the moving generator) to electrical energy?

In this chapter you will learn how these generators work, together with some of the laws of physics which are concerned with generating electricity using magnets.

Fig. 17.2 A small dynamo on a bicycle

17.2 You should know . . .

In order to understand this chapter, there is quite a lot you need to know about magnets and electromagnets. You also need to know how to make simple ammeters and motors. What follows is a brief reminder of this subject. A more detailed explanation can be found in *Physics in Action* Chapter 3.

Magnets have two **poles**—a **north pole** and a **south pole**. The 'like' poles of two magnets repel each other; 'unlike' poles attract. The poles of a magnet are usually, but not always, near the ends of the magnet.

The region round a magnet where magnetic forces act is called a **magnetic field**. Its shape can be shown using iron filings. The iron filings are scattered on a piece of card lying on top of the magnet (Fig. 17.3). A small compass placed in the field shows the direction of the field, which goes from the north pole to the south pole of a magnet.

Fig. 17.4 shows a simple **electromagnet**. The iron core becomes magnetic when a current flows through the wire. Fig. 17.5 shows you how to predict which end of the iron will act as the north pole and which as the south. There is a magnetic field round any electric current. The iron core of an electromagnet makes this field stronger.

If a wire with an electric current flowing in it is in a magnetic field then there is a force on the wire. This can be shown with the apparatus illustrated in Fig. 17.7. The loose piece of wire slides along the other two pieces when the current is switched on. The force causing this movement is at right angles to both the current and the field directions (Fig. 17.8). It is this force which makes a moving-coil ammeter move and some electric motors turn.

Fig. 17.4   A simple electromagnet

Fig. 17.3   The shape of a magnetic field round a bar magnet. What is the plotting compass for?

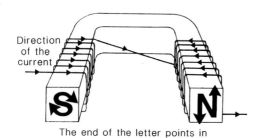

Direction of the current

S   **N**

The end of the letter points in the same direction as the current

Fig. 17.5   How to work out which pole is which for an electromagnet

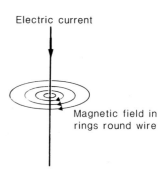

Electric current

Magnetic field in rings round wire

Fig. 17.6   The shape of a magnetic field round a wire carrying an electric current

Fig. 17.7   The 'catapult' experiment

Fig. 17.9   A model moving-coil ammeter

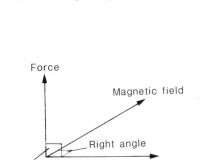

Fig. 17.8   The electric current, the magnetic field
and the force on the wire are all at
right angles to each other

A model moving-coil ammeter is illustrated in the next photograph. Current flows *into* the meter through one spiral spring, round the coil wound on the wooden centre, and out of the other spiral spring. Since the current is flowing along the sides of the coil at right angles to a magnetic field, there will be a force pulling one side of the coil *up*, and the other side *down* (since the current flows in opposite directions along each side). The meter will turn until the force due to the twist of the spiral spring is equal to the electromagnetic force turning the meter.

The **sensitivity** of an ammeter is the amount of movement of the pointer for every ampere of current. How do you think you can change the sensitivity of an ammeter? There are several possibilities.

If possible, find a real moving-coil ammeter in which you can see the mechanism. Try to identify the various parts.

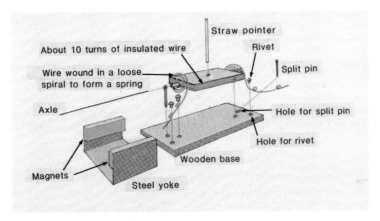

Fig. 17.10   A moving-coil ammeter

17.3 Motors and dynamos

The moving-coil ammeter illustrated in the last section can easily be modified so that the coil keeps turning. The springs are replaced by a **commutator** and two **brushes** (Fig. 17.11). This sliding arrangement allows electric current to pass into and out of the motor without any wires becoming twisted, and also reverses the direction of the current in the coil every half revolution so that the motor keeps turning in the same direction. (See Fig. 17.12 for further details.)

Fig. 17.11   The brushes and commutator of a model electric motor

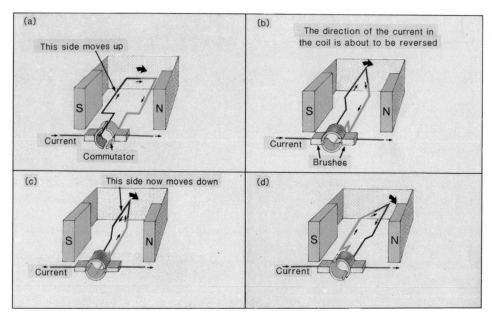

Fig. 17.12   The brushes and commutator ensure that the motor keeps turning in the same direction all the time

Make yourself a model motor, even if you have done so before. Fig. 17.13 should make it clear how to do so. When complete, connect your motor to a power supply of about 4–6 V d.c. (it may need a small push to start it turning).

If your motor does *not* work, check:
1 the magnets have unlike poles facing (i.e. they attract each other);
2 the brushes are touching the commutator.

The torque (twisting force) produced by this motor is not very even; there is a maximum force when the coil is horizontal, and no force when the coil is vertical. (Why not?) To overcome this, real motors have several coils wound on the central **armature**, as it is called, with a corresponding number of segments on the commutator, as shown in Fig. 17.14. Only the coil which will produce the maximum torque is energized at any one time. The greater the number of coils, the smoother running the motor.

The motor you have made can *also* be used as a **dynamo**. In other words, instead of producing kinetic energy from electrical energy, it can produce electrical energy from kinetic energy (Fig. 17.15). To show this happening, make sure your motor is working properly, then switch off and disconnect the power supply. Connect your motor to an ammeter. Wind about 0.5 metre of cotton round the axle, then pull the cotton sharply to rotate the coil (Fig. 17.16). Do you see any movement on the meter to show that a current is flowing?

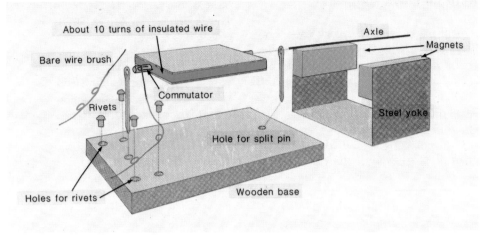

Fig. 17.13   An exploded view of a model motor

Fig. 17.14   The commutator of a motor from an electric drill. Why are there many more coils than on the model motor?

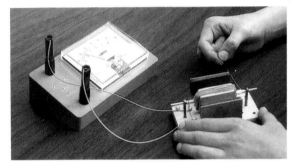

Fig. 17.16   A motor will also act as a dynamo. The coil is being rotated by pulling a piece of cotton wound round the axle

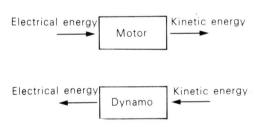

Fig. 17.15

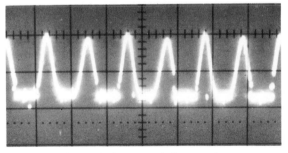

Fig. 17.17   The output from a simple dynamo, shown on an oscilloscope trace. Why does the output vary in the way it does?

You can learn more about the output from your dynamo if you connect the output to an oscilloscope instead. (This is easier to do if you use a device like a VELA to record continuously and store the output from your dynamo while it is rotating, and then display the results on an oscilloscope afterwards.) Fig. 17.17 shows a typical oscilloscope trace of the output from a simple dynamo.

Explain why the output from your dynamo is not steady, but rises and falls in a somewhat irregular way as shown on the oscilloscope trace.

To overcome this uneven output, dynamos may be made with several coils (as was described for motors), with each coil being connected in turn to the brushes when it is producing its maximum output.

It is important to realize that a simple d.c. motor and a simple d.c. dynamo (or generator) can be constructed in exactly the same way. Whether it is called a motor or a dynamo depends on what it is being used for. However, do not imagine that this applies to *all* designs of motors and generators.

17.4 Electromagnetic induction

Using a magnetic field to generate electrical energy in the way you did in the last section is called **electromagnetic induction**. There are three 'ingredients' needed to generate electrical energy in this way:

1. a conductor (e.g. a wire);
2. a magnetic field;
3. relative movement between the conductor and the magnetic field (remember that the dynamo only produced electrical energy when it was turning).

These factors can be investigated in a little more detail using the apparatus shown in Fig. 17.18. The conductor is just a single length of wire, moved by Richard's hands through the magnetic field of the horseshoe magnet. The wire is connected (via a small amplifier) to a VELA, which has been instructed to record the voltage generated across the ends of the wire every 5 milliseconds, and to store these readings so that they can be examined later.

Why do you think it is better to use VELA in this way than to connect the wire directly to an ordinary voltmeter?

Richard passed the wire quickly *down* through the magnetic field, paused a second, brought the wire quickly *up* through the field, paused again, and then repeated these actions slowly. An oscilloscope trace of the results stored in the VELA is displayed in the next photograph.

Look at this photograph, together with Fig. 17.20, and answer these questions.

1. Why is the second peak on the trace in the opposite direction to the first one?
2. Why was there sometimes no p.d. generated?
3. Why are the third and fourth peaks less high than the first two, showing that a smaller p.d. was generated?
4. The third and fourth peaks are *broader* than the first two, i.e. they lasted for a longer time. Why is this?

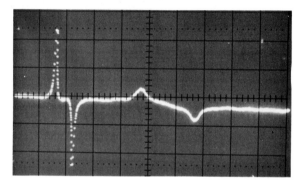

Fig. 17.19   A close view of the oscilloscope trace from Richard's and Sheila's experiment in the last photograph

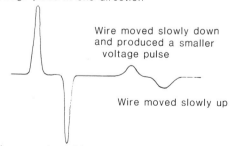

Wire moved quickly down through the magnetic field and produced a large voltage pulse in one direction

Wire moved slowly down and produced a smaller voltage pulse

Wire moved slowly up

Wire moved quickly up through the magnetic field and produced a large voltage pulse in the opposite direction

Fig. 17.18   Investigating the voltage generated when a single wire is moved through a magnetic field

Fig. 17.20   An explanation for the oscilloscope trace in Fig. 17.19

The experiment can be repeated using a much weaker magnet, but with everything else remaining the same. Fig. 17.21 shows the oscilloscope trace of the results. What conclusion do you come to from these results?

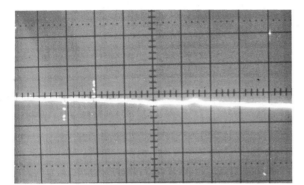

Fig. 17.21 The oscilloscope trace obtained by Sheila and Richard when they used a weaker magnet

Notice that the VELA is measuring the *potential difference* across the wire. Moving a wire through a magnetic field always produces a p.d.; a current will flow only if there is a complete circuit for it to flow along.

A p.d. is only generated if the wire 'cuts' the magnetic field. (Imagine the field lines, or **flux lines**, flowing from the north to the south pole of the magnet; the wire is 'cutting' across them.) Try moving the wire in a line between the north and south poles. Can you generate any electrical energy?

A magnet and a coil, as illustrated in Fig. 17.22, can also be used to investigate electromagnetic induction. The coil can be connected to a 0–100 mV voltmeter as shown, or to an oscilloscope, or to a VELA as before. Set this apparatus up for yourself and then answer the questions below.

1 Do you generate a p.d. when the magnet is inside the coil, but not moving?
2 Is the direction of the p.d. the same whether you move the magnet towards the coil, or take it away from the coil?
3 What is the effect of using different ends of the magnet?
4 Does it matter which end of the coil you use?
5 Does the speed at which you move the magnet towards the coil make any difference to the p.d. you generate?
6 Suppose you use a coil with a different number of turns. Does this affect the p.d. you generate?
7 Do the results from this experiment generally agree with those from the previous experiment?

Fig. 17.22 How to investigate electromagnetic induction using a coil of wire

17.5 The laws of electromagnetic induction

The results of the experiments in the last section can be summarized in the following *laws* (or rules).

Faraday's law of electromagnetic induction

The potential difference generated across a conductor moving in a magnetic field is proportional to the rate at which the conductor cuts through the magnetic field lines.

Note: the 'rate of cutting magnetic field lines' (and hence the p.d. induced) can be increased by:
1 having a *stronger magnet* so that there are more field lines;
2 moving *faster* through the field;
3 moving *more conductor* through the field.

Which experiments in the last section showed that each of these factors do have an effect on the p.d. generated?

197

Fleming's right-hand rule

If the thumb and first two fingers of the right *hand are held at right angles to each other as in Fig. 17.23, the second finger points in the direction of the current if the first finger points in the direction of the magnetic field and the thumb points in the direction of the motion of the wire.*

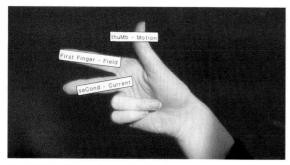

Fig. 17.23   Fleming's right-hand rule

If you *have* to remember this, the following might help:

thu<u>M</u>b — <u>M</u>otion
<u>F</u>irst <u>F</u>inger — <u>F</u>ield
se<u>C</u>ond finger — <u>C</u>urrent

Lenz's law

An induced current always flows in such a direction as to oppose the change producing it.

Look at Fig. 17.24. As the north pole of the magnet approaches the coil, the current in the coil flows in such a way as to produce a north pole at the end of the coil nearer the magnet, so as to oppose the movement of the magnet. If you look back at Fig. 17.5 you will see that the current must therefore flow in the direction shown.

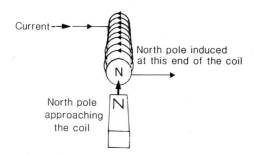

Fig. 17.24   How to decide which way the current is induced in a coil

Lenz's law really follows from the law of conservation of energy. If a *south* pole were induced in the coil in Fig. 17.24, this would attract the magnet, and the magnet could then move up to the coil all by itself, generating more electrical energy as it did so. This is not allowed!

17.6   The a.c. generator

The generator that you made in Section 17.3 made *direct current*—current which flows the same way all the time. Which part of the generator is responsible for ensuring that the current *does* flow the same way all the time?

Many generators, especially large ones such as are in the power station in Fig. 17.1, produce *alternating current*—such generators are sometimes called **alternators**. One reason for producing a.c. at power stations is that it can easily be transformed to a higher or a lower voltage, and it is essential to be able to do this if electricity is to be transmitted over large distances, as you will see in the next chapter.

It is easy to modify the generator you made in Section 17.3 to produce a.c. Each end of the wire forming the coil is wrapped round one end of the axle as in Fig. 17.25, so that each brush is always connected to the same end of the coil. Look at Fig. 17.26, which is a simplified diagram of the generator. As the *red* side comes up through the field, the current flows in the direction shown and out of end B; later, as the red side moves *down* through the field, the current flows the opposite way and out of end A. A graph of current against time will look like Fig. 17.27.

Fig. 17.25   A model of a simple a.c. generator

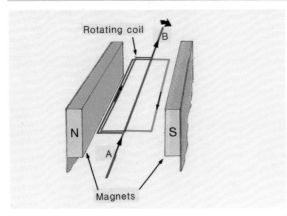

Fig. 17.26   A simple a.c. generator

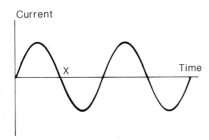

Fig. 17.27   A graph of current against time for a simple a.c. generator

Why is there *no* current at the point marked X, even though the generator is still turning so the wire is still moving in the magnetic field?

Make a generator like the one illustrated in Fig. 17.25 and check for yourself that it does give a.c. by connecting the output to a meter, oscilloscope or a VELA in the same way as you did before.

In practice, an a.c. generator of the kind you have made will usually have its brushes and **slip rings**, as they are called, at the same end of the axle using an arrangement like the one in Fig. 17.28.

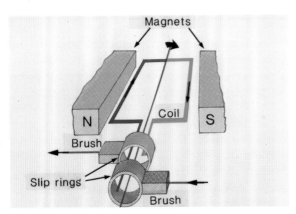

Fig. 17.28   The slip rings and brushes of an a.c. generator

However, there is a much simpler way of generating a.c. than using a generator like those so far illustrated. It does not matter which moves, the coil or the magnet, so long as one moves relative to the other. Would it not be much simpler to rotate a *magnet*, and keep the coil stationary?

Fig. 17.29 shows a very simple arrangement with a magnet rotating near the end of a coil. It can be improved in the way shown in the next photograph, using two coils wired in series with the magnet rotating between them. Try this arrangement for yourself if possible. If you use a moving-coil meter to detect the generated electricity, you will soon discover that, as you rotate the magnet faster, the inertia of the needle prevents it from keeping pace with the rapid fluctuations in the current. The photographs suggest other ways to detect the rapidly alternating voltages.

Fig. 17.29   A model rotating magnet generator

The arrangement in Fig. 17.29 can be further improved by placing the coils on an iron core. Why is this an improvement?

Fig. 17.30   The generator in the previous photograph can be improved by using two coils

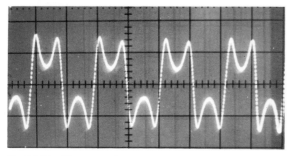

Fig. 17.31   The output from a rotating magnet generator shown on an oscilloscope

The bicycle generator illustrated in Fig. 17.2 is made in this way with a magnet rotating between a pair of coils. Why do you think this is more reliable than rotating a coil in a magnetic field in the way we did at the start of this section?

17.7   Generators in power stations

There is no reason why there should not be more than one pair of coils round the rotating magnet in Fig. 17.29. For example, the rotating magnet in Fig. 17.32 will generate a p.d. in all three pairs of coils. The output from each pair of coils will be 'out of phase' with its neighbours, as illustrated in Fig. 17.33.

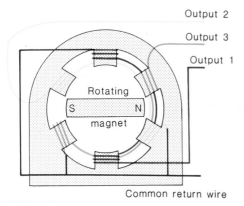

Fig. 17.32   A three-phase generator

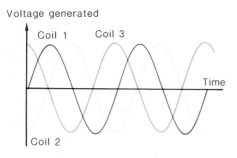

Fig. 17.33   The voltage output from a three-phase generator

A generator like this is called a **three-phase generator** and is the sort of generator found in power stations. Fig. 17.34 shows one such generator being assembled. In this case the rotating magnet is an *electromagnet*, so a brush and commutator is needed to pass this current into the electromagnet. However, the current needed for this electromagnet is small compared to the current being generated, so the brushes and commutator for the electromagnet are easier to construct than brushes and slip rings for rotating coils. Also, six brushes and six slip rings would be needed. Why?

Fig. 17.34   Assembling a large generator

The current for the electromagnet is usually obtained from a small d.c. generator driven by the same turbine as the main generator. This small d.c. generator is sometimes called an **exciter**. Why do you think it is given this name?

The *force* needed to turn the magnet of an a.c. generator varies as it turns. A large force is needed when it is near the coils and a lot of electrical energy is being generated [Fig. 17.35(a)]. While in the position shown in Fig. 17.35(b) a much smaller force is needed because very little electrical energy is being generated at this moment. With a large single-phase generator this varying force can give rise to vibrations, which will cause rapid wear in

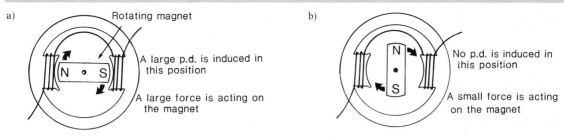

Fig. 17.35   Why does the force required to drive a generator vary with the position of the coil?

the generator. Why do you think this is not such a problem with a three-phase generator?

You should now understand all the stages in obtaining electrical energy from coal or oil. The electricity still has to get from the power station to your home. You will learn how this is done in the next chapter.

SUMMARY

Now that you have finished studying this chapter on generating electricity, there are a number of things you should know and be able to do.

1 You should:
   a) be able to make a model dynamo;
   b) know Faraday's law of electromagnetic induction;
   c) know Fleming's right-hand rule;
   d) know Lenz's law;
   e) understand why motors and dynamos usually have more than one coil;
   f) be able to explain how an a.c. generator works.

2 You should know what each of the following is, or what each does:

sensitivity (of a meter)	brush
	exciter
commutator	alternator
armature	slip rings
flux lines	

3 These are some of the other words that have been used in this chapter. You should know what each word means:

scattered	region
mechanism	identify
fluctuation	

FURTHER QUESTIONS

1 A magnet is dropped through a coil of wire, as in Fig. 17.36
   a) How do you think the p.d. across the ends of the coil will vary as the magnet is dropped through the coil?
   b) Design an experiment to test whether or not your answer is correct.

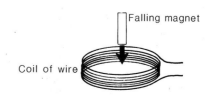

Fig. 17.36

2 A long magnet is suspended from a spring so that one pole is level with the end of a coil of wire as illustrated in Fig. 17.37. (The other pole is a long way from the end of the coil.) The magnet is made to oscillate up and down on the end of the spring.

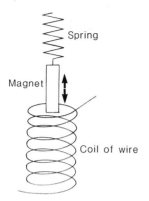

Fig. 17.37

a) Explain how the p.d. generated across the ends of the coil will vary as the magnet oscillates on the end of the spring.
b) At what part of the magnet's oscillation will the greatest p.d. be generated?
c) What difference would it make to the p.d. generated across the coil if a *short* magnet were to be suspended in the *middle* of the coil and made to oscillate?

3 An a.c. generator is connected to an oscilloscope. When the generator is turning, the oscilloscope trace is as illustrated in Fig. 17.38. What will be the effect on the trace if:

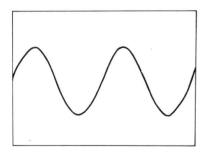

Fig. 17.38

a) the generator's magnets are replaced with stronger magnets;
b) the original magnets are used, but the generator is turned twice as fast as before;
c) the original magnets are used, and the generator is turned at the original speed, but twice as many turns of wire are used in making the generator?

4 Fig. 17.39 shows a coil of wire connected by a centre-zero voltmeter which has a high resistance. The coil is free to swing between the poles of a large magnet as shown.

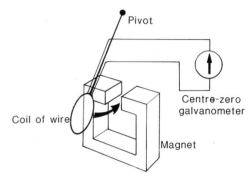

Fig. 17.39

a) Describe and explain how the voltmeter reading changes during one complete swing of the coil.
b) If the terminals of the voltmeter are connected together with a length of wire, thus 'shorting' the coil, the coil stops swinging very quickly. Explain, in terms of energy, why this is so.

5 In Britain, the Earth's magnetic field lines are at an angle of about 70° to the horizontal. The handlebars of a bicycle moving along a level road will cut through these field lines, and thus a p.d. should be generated across the ends of the handlebars. Explain whether or not it should be possible to use suitable lengths of wire to connect the ends of the handlebars to a sensitive voltmeter mounted in the centre of the handlebars to detect this p.d.

TRANSFORMERS AND POWER LINES

Fig. 18.1   This large substation, which forms part of the national grid, uses transformers to change electricity from one voltage to another

18.1   Introduction

The first thing that happens to the electricity as it leaves the power station is that a transformer like one of those in the photograph changes it to a much higher voltage.

Electricity is generally generated at about 25 kV, but it is distributed at potentials as high as 400 kV. It is then transformed down to lower voltages for use in factories, schools and homes.

In this chapter you will learn how a transformer changes the voltage, and why it is necessary to do so.

18.2   You should know . . .

In order to understand this chapter, you should:
1 have read and understood all of the previous chapter;
2 know and understand how to calculate electric power and resistance.

18.3   Induction coils

You can begin to understand how a transformer works by first looking at something which produces the high voltage needed for the sparking plugs in a petrol engine.

In Section 17.4 you saw that a p.d. is generated across a coil if a magnet moves towards or away from the coil. Of course, an electromagnet can be used in place of a permanent magnet (Fig. 18.2), but there is another way of using an electromagnet.

Fig. 18.2   A moving electromagnet will generate a voltage in a coil of wire. Compare this illustration with Fig. 17.22

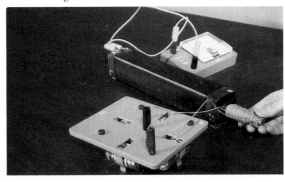

Fig. 18.3 How can a voltage be induced in the coil without moving the electromagnet?

Place an electromagnet next to a coil, and connect the coil to a voltmeter, as shown in Fig. 18.3. Switch on the current to the electromagnet. Just for an instant, while the magnetic field round the electromagnet is changing, a p.d. is **induced** across the coil. Did your voltmeter give a reading just for a moment? If the electromagnet is now switched off, a p.d. is induced in the coil for just a moment in the opposite direction.

Notice that in these small experiments nothing has *moved* in the way that the magnet moved in Section 17.4. So long as the magnetic field is *changing in strength*, a p.d. will be induced across the coil.

Quite a high voltage can be induced in this way. In the experiment shown in Fig. 18.4, a neon lamp is connected across the coil. A neon lamp consists of two electrodes in a glass envelope filled with neon gas. When there is a p.d. of about 90 V across the neon it glows red. Switching the magnetic field on or off generates enough p.d. across the coil to light the neon lamp, but only, of course, for a moment while the field is actually changing. Take care if you try this yourself. You will certainly feel a 90 V shock!

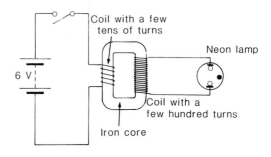

Fig. 18.4 How can this circuit generate a sufficiently high voltage to light the neon bulb?

This idea is used to generate the high voltage to the sparking plugs of a petrol engine. You will know how a petrol engine works if you have studied Section 8.6. The complete electrical system for producing the spark is illustrated in Fig. 18.5. The **induction coil** (Fig. 18.6) is supplied with current via a switch, or **contact breaker** (Fig. 18.7) which is opened and closed by a cam rotated by the engine. (What is a 'cam'?)

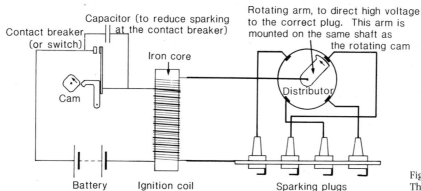

Fig. 18.5
The ignition system of a car

Fig. 18.6   An induction coil in a car. How does it generate the high voltage for the sparking plugs?

Fig. 18.7   The contact breaker

As the switch is opened (or closed) a large p.d. (over $10\,kV$) is generated across the **secondary winding** of the induction coil. This large p.d. is routed, via well insulated leads, to the correct sparking plug by a rotating arm on top of the contact breaker assembly. The contact breaker opens at just the right moment for a spark to appear at the plug when the fuel needs igniting. You can see the adjustment screw on the contact breaker assembly in Fig. 18.7 that enables this 'timing' to be adjusted accurately.

What do you think happens if the high voltage leads are *not* well insulated? Suppose they become damp, for example?

18.4   Transformers

Remember that a p.d. is only generated in the coil in Fig. 18.3 when the magnetic field is *changing*. If it were possible for the magnetic field to change all the time, then a p.d. would be generated all the time. But this *is* possible—by using an alternating current in the electromagnet (see Fig. 18.8). Notice that what we have so far called the electromagnet is now called the **primary coil** in this arrangement. Alternating current flows through the primary coil, so making a *changing magnetic field* in the iron core. A potential difference is induced in the **secondary coil** because the changing magnetic field passes through this coil. This arrangement is called a **transformer**.

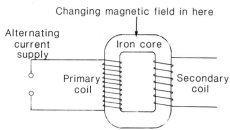

Changing magnetic field in here

Alternating current supply

Iron core

Primary coil

Secondary coil

The current in the primary coil causes a changing magnetic field in the iron core

A p.d. is generated in the secondary coil because it is in a changing magnetic field

Fig. 18.8   A simple transformer

Transformers can have the p.d. across the secondary *less* than that across the primary (a **step-down transformer**), or with the secondary p.d. *greater* than the primary p.d. (a **step-up transformer**). The electricity distribution system in this country uses both step-up transformers, to increase the p.d. generated at the power station for distribution, and step-down transformers to decrease this p.d. to a suitable value for use in houses and factories.

There is a simple way of calculating the p.d. across the output (secondary) coil of a transformer. The equation is:

$$\frac{\text{p.d. across secondary}}{\text{p.d. across primary}} = \frac{\text{no. of turns on secondary}}{\text{no. of turns on primary}}$$

You can easily *test* this equation using apparatus similar to that in Fig. 18.9. Choose one coil, for example, an 800 turn coil, to be the primary coil, and connect this to a 4 V a.c. supply. Use other coils in turn as secondary coils and measure the p.d. across each secondary coil. The photograph does not show an instrument for measuring this p.d. across the secondary, nor for checking that the p.d. across the primary is 4 V. You will have to decide for yourself how to do this.

You can obtain a greater number of results by using a number of different coils for the primary as well.

Fig. 18.10   A typical small transformer

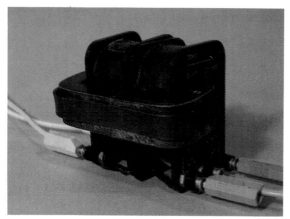

Fig. 18.9   This apparatus can be used to investigate the input and output voltages of a simple transformer

Record your results in a table like this:

Number of turns on primary coil		
Number of turns on secondary coil		
p.d. across primary coil (V)		
p.d. across secondary coil (V)		

Do your results confirm the equation above?

The equation does assume that *all* the magnetic flux generated by the primary coil goes through the secondary coil. With the simple design of transformer using the apparatus in Fig. 18.9 this will not be true because there will be some flux 'leakage'. Practical transformers are carefully designed to reduce this leakage to a minimum. Fig. 18.11 shows how a small step-down transformer is constructed, with the core made up of many layers, or **laminations**, of iron.

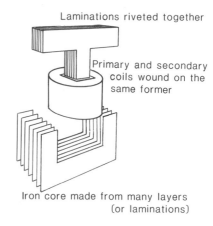

Laminations riveted together

Primary and secondary coils wound on the same former

Iron core made from many layers (or laminations)

Fig. 18.11   The details of how the transformer in the previous photograph is made

18.5   Transformers and energy

The secondary coil of a transformer is supposed to be connected to a circuit so that current can be driven round this circuit. In the experiment in the last section, you measured the output from the secondary coil when it was not driving any current into a 'load'. The output from the secondary coil will *decrease* if a current is flowing through the secondary coil (and round an external circuit). The larger the current the smaller the p.d. This is because some energy is needed to drive the current through the secondary coil itself, since this coil has some resistance (see Fig. 18.12). What happens to this energy? Where have you met a very similar idea to this already?

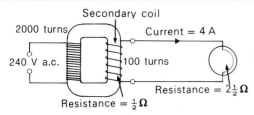

12 V generated in the secondary coil but 2 V is needed to drive 4 A through $\frac{1}{2}\,\Omega$ resistance therefore there are 10 V across the terminals of the transformer

Fig. 18.12   Some energy is needed to drive current through the secondary coils

It may have occurred to you that a step-up transformer appeared to disobey the law of conservation of energy. You get more joules per coulomb *out* than you put *in*. However, you also get fewer coulombs per second, so you do not get more energy per second (i.e. power) out than you put in. In a 100% efficient transformer, the output power (volts × amps) will be equal to the input power, so if the output voltage is *larger* than the input voltage, then the output current will be *smaller* than the input current. Try to modify your last experiment to test the truth of this.

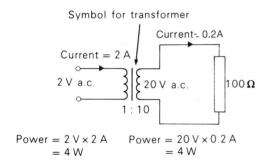

Input power = output power

Fig. 18.13   A transformer obeys the law of conservation of energy

Many practical transformers such as the one illustrated in Fig. 18.10, have more than one secondary coil on the same core. Since each secondary coil is in a changing magnetic field, a p.d. will be generated across each secondary coil.

Another variation is to tap-off the secondary coil at several points. Such a transformer may well be inside your laboratory power supply. A special case of this is a **centre-tapped transformer**, which is particularly useful for providing a full-wave rectified output using the circuit shown in Fig. 18.14.

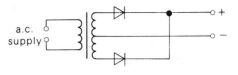

Centre-tapped transformer

Fig. 18.14   A centre-tapped transformer being used with a rectifier to convert a.c. to 'full wave rectified' d.c.

Try to work out how this circuit functions. What will the graph of output p.d. against time from this circuit look like?

18.6   A numerical example

This example will help you understand the ideas in the previous section.

A transformer has a primary coil and two secondary coils as shown in Fig. 18.15. The primary coil is connected to the 240 V mains.

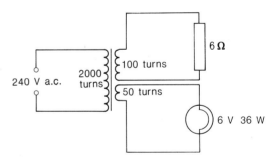

Fig. 18.15

a) What is the p.d. across each secondary?
b) Assuming each secondary has negligible resistance, how much current flows in each secondary circuit?
c) What is the total power output from the transformer?
d) What is the power input to the transformer?
e) What assumption do you have to make to answer part e)?
f) What is the input current?

Part a)
Using
$$\frac{V_s}{V_p} = \frac{N_s}{N_p}$$

$$V_s = \frac{N_s}{N_p} \times V_p$$

For the 100 turn coil: For the 50 turn coil:

$$V_s = \frac{100}{2000} \times 240\,V \qquad V_s = \frac{50}{2000} \times 240\,V$$

$$= 12\,V \qquad\qquad\qquad = 6\,V$$

Part b)
For the 100 turn coil, we can use the relationship $I = V/R$ to calculate current:

$$I = \frac{12\,V}{6\,\Omega}$$

$$= 2\,A$$

For the 50 turn coil, we can use the relationship $I = P/V$ to calculate current

$$I = \frac{36\,W}{6\,V}$$

$$= 6\,A$$

Part c)
The power output from the 100 turn coil can be calculated from

$$P = V \times I$$
$$= 12\,V \times 2\,A$$
$$= 24\,W$$

We already know the power output from the 50 turn coil is 36 W, so the total power output is 24 W + 36 W = 60 W.

Part d)
According to the law of conservation of energy, the power input to the transformer will be the same as the power output, that is, 60 W.

Part e)
We assume there are no energy losses; e.g. the wires and the core do not heat up.

Part f)
Using
$$I = \frac{P}{V} \qquad = \frac{60\,W}{240\,V}$$

$$= 0.25\,A$$

The input current is 0.25 A.

18.7 Why is a transformer core laminated?

In Section 18.4 you saw that the core of a transformer is made from *layers* of iron, i.e. it is **laminated**. The iron cores in the photographs in this chapter are laminated. Why do they have layers? Why not use a solid block of iron? Are the iron cores you have used laminated?

Between each layer of iron there is a layer of varnish. This construction reduces electric currents which are induced in the iron core itself. The iron core is a conductor in a changing magnetic field, so there will be a p.d. induced in the core. Since a solid block of iron has a low resistance, a large current would flow round a solid core as illustrated in Fig. 18.16. What do you think this large current would do to the core?

To reduce these **eddy currents**, as they are called, the core is made from layers, insulated from each other—see Fig. 18.17. This does not stop eddy currents altogether, but it does reduce them to a very small level. Even so, a large transformer must be cooled with circulating oil. The iron core in electric motors is laminated for exactly the same reason.

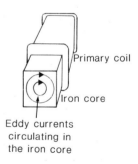

Eddy currents
circulating in
the iron core

Fig. 18.16 'Eddy currents' flowing in a transformer core

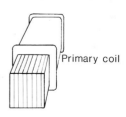

Iron core made of layers.
This reduces the eddy currents

Fig. 18.17 Reducing eddy currents by 'laminating' the core

Fig. 18.18 shows another example of eddy currents at work. The aluminium trough is behaving as a single-turn secondary winding of a transformer. The large coil at the back is supplied from the mains, and a p.d. is induced in the trough (it is a small p.d.).

(What p.d. will be induced if the primary coil has 800 turns and is supplied from the 240 V mains?)

Fig. 18.18 Eddy currents induced in this small trough make it hot enough to boil the water

However, the resistance of the coil is very low, so a large current can flow (about 300 A). The trough soon becomes hot—it took less than half a minute for the water in the trough to start boiling.

Domestic cookers which use this idea are becoming available, with large currents being induced in the saucepans to heat them. Why will these electric currents not be dangerous? Do you think such a cooker has any advantages over a conventional cooker? What might happen to any rings or bracelets that the cook was wearing?

Large induction furnaces have been used for many years in the steel industry to melt iron. The iron is placed in the furnace in which there is a large changing magnetic field. The eddy currents induced in the iron make it hot enough to melt.

18.8 Power lines

We can now turn to the question of why the Electricity Generating Board transmits electrical energy round the country at a high voltage, and then steps it down, using transformers, to 240 V for supply to your home. Would it not be cheaper to generate the power at 240 V in the first place?

We can investigate the reason for this on a laboratory scale. Suppose there is a 'power station' in the 'town' of Exe generating power at 12 V. (An ordinary lab power supply will serve as the power station.) A 12 V, 36 W lamp in the town connected to this supply lights up perfectly satisfactorily, as you can see in Fig. 18.19.

Fig. 18.19 Lighting a lamp at Exe

Imagine we now want to connect the 'village' of Wye, further down the laboratory, to the same power station. We connect the power station to the village with a pair of cables, or 'power line', and connect a second 12 V, 36 W lamp to the power line in the village. Fig. 18.20 shows the result—the lamp only lights dimly!

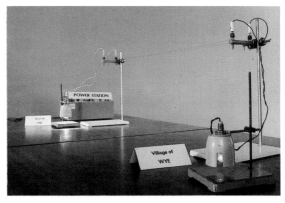

Fig. 18.20 The lamp at Wye is not nearly as bright as the one at Exe

If we repeat the experiment using a high voltage, e.g. the 240 V mains with a 240 V, 40 W lamp (i.e. about the same power as before), then there is no problem in lighting the lamp in the village (Fig. 18.21). However, this is now much more dangerous than it was before, and, for this reason, *do not do this experiment yourself*. Let your teacher demonstrate it to you.

Fig. 18.21   When mains bulbs are used, they are as bright as each other

Exactly the same problem applies on a larger scale. The energy losses are smallest if the electricity is transmitted at a very high voltage—up to 400 kV in Britain. This is certainly not safe to bring into your house.

The final solution, then, is to use a high voltage for *transmitting* the energy and to reduce the voltage to a lower, safer value where the electricity is needed. Applying this idea to our model system, Fig. 18.22 shows the final design. Electricity is 'generated' at 12 V and stepped-up to 240 V using a 20 : 1 step-up transformer. At Wye it is stepped-down to 12 V again and you can see the lamp lights perfectly satisfactorily this time.

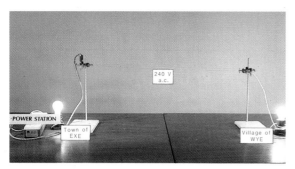

Fig. 18.22   Using a power line with transformers at each end

Why does using a high voltage reduce the energy loss in the power line? The cause of the problem is the resistance of the power line cables. Each cable in our model in Fig. 18.20 happened to have a resistance of 2 ohms. Each 12 V, 36 W lamp has a resistance of 4 ohms. (How has this been calculated?) A circuit diagram is shown in Fig. 18.23.

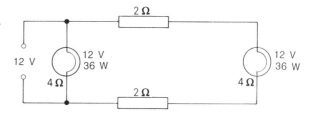

Fig. 18.23

1 How much current flows through the lamp at Exe?
2 What is the *total* resistance of the circuit through Wye?
3 How much current flows through the circuit to Wye?
4 What is the potential difference across the lamp at Wye?

You should be able to see (if you have answered these questions correctly) that half of the energy supplied to Wye is lost in the cables, warming them up.

Now let us repeat these calculations for the 240 V supply. A 240 V, 40 W lamp is designed to have a current of $\frac{1}{6}$ A through it, so the resistance of the filament is

$$\frac{240\,\text{V}}{1/6\,\text{A}} = 1440\,\Omega$$

The extra 4 ohms cable resistance is negligible compared to 1440 ohms, so the current going to Wye will be about $\frac{1}{6}$ A. The voltage drop down each cable is given by

$$\begin{aligned} V &= I \times R \\ &= \tfrac{1}{6} \times 2\,\text{V} \\ &= \tfrac{1}{3}\,\text{V} \end{aligned}$$

The total for both cables is $\frac{2}{3}$ V. This is negligible compared to the 240 V we started with, so virtually all this 240 V is available at Wye.

To summarize: for a given power, the *larger* the voltage, the *smaller* the current through the cable. A small current does not need much energy to push it through the cable, so the energy losses are small and the voltage drop down the cable is small.

A diagram of the complete electricity generating and transmission system appears in Fig. 18.24. Make sure you know why each part is there and how it works.

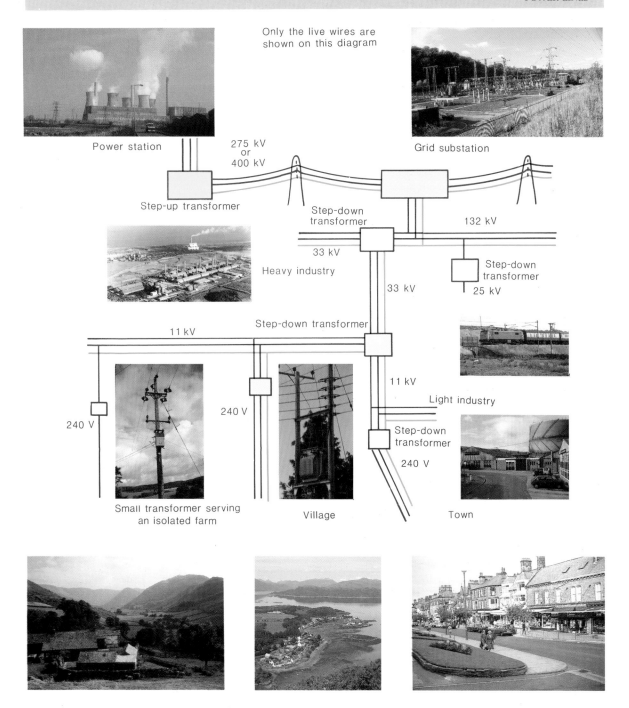

Only the live wires are shown on this diagram

Power station

275 kV or 400 kV

Grid substation

Step-up transformer

Step-down transformer

132 kV

33 kV

Heavy industry

Step-down transformer

25 kV

33 kV

Step-down transformer

11 kV

11 kV

Light industry

240 V

240 V

Step-down transformer

240 V

Small transformer serving an isolated farm

Village

Town

Fig. 18.24   A simplified diagram of the national grid system

SUMMARY

Now that you have finished studying this chapter on transformers and power lines, there are a number of things you should know and be able to do.

1 You should:

a) be able to describe the construction of a transformer, and explain how it works;

b) know how a high voltage is generated for the sparking plugs of a petrol engine;

c) know that a p.d. can be induced if a magnetic field *changes*;

d) know how to calculate the output voltage from a transformer;

e) be able to apply the law of conservation of energy to transformers;

f) understand why a transformer core is made of layers;

g) know how an induction furnace works;

h) understand why electric power is transmitted over long distances at a high voltage.

2 You should know what each of the following is, or what each does:

primary coil	secondary coil
contact breaker	lamination
tapping	induction coil
centre-tapped	cam
transformer	eddy current

3 These are some of the other words that have been used in this chapter. You should know what each word means:

distribution	ignite
circulating	domestic

FURTHER QUESTIONS

1 Fig. 18.25 shows a cell and a variable resistor connected via a switch to a coil of wire wound round an iron bar. Also wound round the bar is another coil of wire connected to a galvanometer. If the switch is closed, so that current starts to flow in the first coil, the galvanometer momentarily moves to the *right* and then returns to its central zero position.

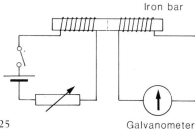

Fig. 18.25   Galvanometer

a) Explain why the galvanometer only moves momentarily.

Describe and explain what happens to the galvanometer in each of the following cases.

b) The switch is opened again so that current ceases to flow in the first coil.

c) The switch is closed again, and then the variable resistor is slowly reduced to zero resistance.

d) Whilst the switch is still closed, the iron bar is broken along the dotted line.

e) The iron bar is mended again and the cell is replaced with an alternating current supply, the switch still being closed.

2 Fig. 18.26 shows different arrangements of the same transformer. P is a primary coil of 20 turns, which is connected to a 1 V a.c. supply. S is a secondary coil of 50 turns joined to a 2.5 V lamp. In arrangement (c) the secondary is wound on top of the primary.

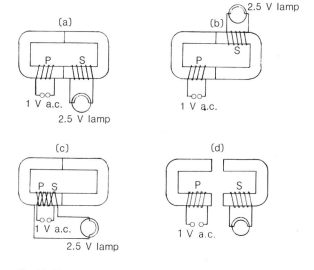

Fig. 18.26

a) There seems to be no difference between arrangements (a), (b) and (c). In each case the lamp lights normally although the secondary is in different places. How do you explain this?

b) Diagram (d) shows the same arrangement as (a), but the two halves of the iron core have been slightly separated. This time the lamp only lights very dimly. How do you explain this?

c) If all three secondaries from diagrams (a), (b) and (c) are placed on one transformer core with one primary, as in Fig. 18.27, and each secondary is joined to a 2.5 V lamp, each lamp is found to light normally. How do you explain this result?

d) How would you expect the primary current in the arrangement in Fig. 18.27 to compare with the primary current in the arrangement in Fig. 18.26(a)?

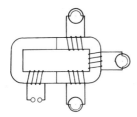

Fig. 18.27

3 If a transformer, such as you may have made, has loose parts, then a hum is heard while there is current flowing through the primary coil. How do you explain this noise?

4 Fig. 18.28(a) shows a transformer with a primary coil connected to an a.c. supply and two secondary coils, each of which gives an output of 6 V.

a) In an attempt to light a 12 V lamp, a pupil joined the two secondary coils together as shown in diagram (b). The lamp did not light at all. Why not?

b) Is there any way in which the two secondary coils could be connected so as to give a 12 V output and so light the lamp properly?

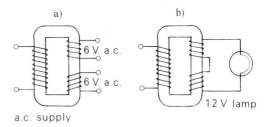

Fig. 18.28

5 A 24 V, 36 W lamp is to be lit using a suitable step-down transformer supplied by the 240 V mains.

a) If there are 1000 turns of the primary coil of the transformer, how many are needed on the secondary coil?

b) If the transformer is 100% efficient, what current is taken from the mains?

6 A small power station generates 5 MW of electric power at a voltage of 10 kV. This power is delivered to a small town some distance away along a power line of total resistance 10 Ω, as shown in Fig. 18.29.

a) What is the current generated by the power station?

b) What p.d. is required to drive this current along the power line?

c) What is the power dissipated in the power line?

d) What fraction of the total power generated is 'lost' in the power line?

e) What happens to this 'lost' power?

f) Explain how you could use suitable transformers to reduce the power loss in the power line.

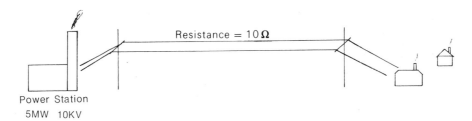

Fig. 18.29

213

19

STATIC ELECTRICITY

Fig. 19.1   What makes lightning?

19.1   Introduction

Early man may well have thought that the vivid lightning flash, and the roar of thunder which followed it, were the anger of the gods showing itself. In fact, lightning is a giant electric spark, as Benjamin Franklin (Who was he?) showed by flying a kite into a thundercloud. Some of the electricity ran down the wet string of the kite into a brass key hung on the end. He was able to draw a small spark from the key, and obtain a small shock, by placing his knuckle near the key.

In fact, this was a very dangerous experiment. A thundercloud stores a very large quantity of electrical energy—think of the damage lightning can do. Franklin was lucky to come to no harm; other people have been killed attempting to do the same experiment. His experiment led him to invent the lightning conductor.

How is electricity stored in the cloud? Is it stored as you store water in a tank? If so, can you store electricity in a similar way in the laboratory? Questions like these belong to that part of physics known as **static electricity**. (What does the word 'static' mean?)

Fig. 19.2   A tree split in two by lightning

19.2   You should know . . .

In order to understand this chapter, you should have read and understood Chapter 14 on electrical energy.

19.3   Capacitors

The object which looks like a small can in Fig. 19.3 is able to store electricity. It is called a **capacitor**. Connect a capacitor like the one in the photograph to a 12 V battery—electric charge will flow into the capacitor. Remove the leads from the battery and hold the plugs very close to each other without letting them touch, as in Fig. 19.4. A tiny spark will jump across the gap—a very small scale lightning flash. Alternatively, using a switch as in the next photograph, connect the capacitor first to a 6 V battery and then to a 6 V, 0.04 A lamp. The lamp lights, but only for a moment. The amount of charge that can be stored is very small (it stores 0.06 coulomb in this experiment) and the capacitor soon *discharges*.

The circuit diagram for this experiment appears in Fig. 19.6. Notice the symbol for the capacitor. This particular capacitor has positive and negative terminals and must be connected the correct way round (it is said to be **polarized**). Other capacitors, of which there are some examples in Fig. 19.7, are not polarized, and the symbol for these is very slightly different (Fig. 19.8).

Fig. 19.3
A capacitor

Fig. 19.4
Vicki making a small spark from a discharging capacitor

Fig. 19.7   Capacitors come in many shapes and sizes

Fig. 19.5   How to discharge a capacitor through a lamp

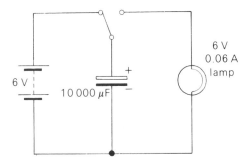

Fig. 19.6   A diagram of the circuit being used in the previous photograph

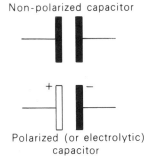

Non-polarized capacitor

Polarized (or electrolytic) capacitor

Fig. 19.8
The symbol for a capacitor

STATIC ELECTRICITY

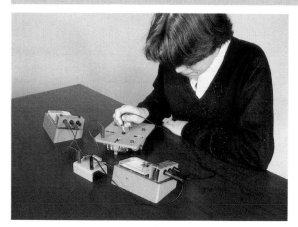

Fig. 19.9   Investigating the current flowing when a capacitor is charged

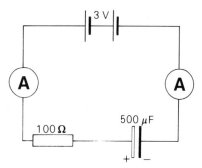

Fig. 19.10   The circuit diagram for the previous photograph

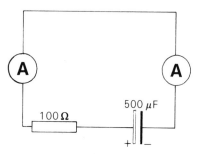

Fig. 19.11   Discharging a capacitor

It is tempting to think that electricity is pouring out of the wires (pushed by the battery) into the capacitor, like water from a tap into a pan. However, the next experiment shows that this analogy is not a very good one.

Connect the circuit illustrated in Fig. 19.10. The 3 V battery charges the capacitor through a 100 ohm resistor to slow down the flow of charge so that you have a chance to see what is happening. Do the following, answering the questions as you proceed.

1 Switch the circuit on. Does any current flow through either ammeter? Does a current keep flowing?

2 Remove the battery and connect the two ends of the circuit together with a connecting lead (Fig. 19.11). Does any current flow in any part of the circuit?

3 Repeat both these experiments using a 6 V battery instead of the 3 V battery. Does the same current flow as before?

Your first experiment may give you the impression that charge is actually flowing *through* the capacitor, but this is not so. A capacitor consists of two plates of metal with an insulator between them (Fig. 19.12). The capacitor shown in Fig. 19.3 has the two plates rolled up, like a swiss roll, but they are still insulated from each other, so electricity cannot flow *through* the capacitor. When a battery is connected across the capacitor, electrons flow *onto* one plate and the same number of electrons flow *off* the other plate, making it positively charged.

Now explain the result of your second experiment using the idea of electrons flowing.

Explain why the 'water tank' analogy is not a very good one for a capacitor.

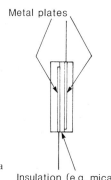

Fig. 19.12
The construction of a simple capacitor

Fig. 19.13
The construction of an electrolytic capacitor

Anode covered in a layer of aluminium oxide

Paper soaked in electrolyte

Aluminium foil

Metal plates

Insulation (e.g. mica)

19.4 A use for a capacitor

If you have read Chapter 16, you will remember that diodes can be used to *rectify* alternating currents, but that the direct current that is produced is not *smooth*. A capacitor can be used to smooth out the variations in this direct current.

Look at Fig. 19.14. When there is a peak in the output from the rectifying circuit, charge flows onto the capacitor so charging it up. At those times when there is little or no output from the rectifying circuit, charge can flow off the capacitor, thus keeping current flowing.

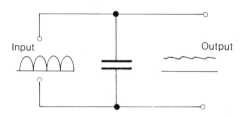

Fig. 19.14 A capacitor can be used to smooth the irregularities in the direct current obtained from a rectifier

The capacitor must be able to store sufficient charge to meet the demand for current when there is no output from the rectifying circuit. The bigger the current demand, the bigger the required capacitor.

A capacitor used in this way is sometimes called a **reservoir capacitor**. Do you think this is a good name for it? Is it doing a similar thing to a water reservoir?

19.5 Capacitance

Fig. 19.15
Is electric charge stored in a capacitor like water stored in a beaker?

How much electric charge can a capacitor hold? The third experiment you did in the last section should have shown that this is not such an easy question to answer as 'How much water does the beaker in the photograph hold?' (Answer: 250 ml).

Once there are 250 ml of water in the beaker, it is full and you cannot put any more water in. With a capacitor, the higher the voltage you connect across the capacitor, the more the charge you can 'push' into it. In practice, there is a limit to the voltage you can put across a capacitor without it breaking down. (Notice that we still talk about 'charging up a capacitor' even though the electrons go *in* one side and *out* of the other.)

To measure the capacitance of a capacitor, we measure the charge stored in the capacitor for every 1 V of potential difference across the capacitor, i.e.

$$\text{capacitance} = \frac{\text{coulombs}}{\text{volts}}$$

The unit of capacitance is the **farad**. A 1 F capacitor will store a charge of 1 C if there is a p.d. of 1 V across its terminals. A 1 F capacitor would be very large indeed. The capacitor shown in Fig. 19.3 has a capacitance of 10 mF, i.e. it stores 10 mC of charge for every volt, and this is one of the largest capacitors you will meet. Capacitors commonly have values measured in microfarads, nanofarads or picofarads (10^{-6}, 10^{-9} or 10^{-12} F).

19.6 Charging by rubbing

A thundercloud is like a very large capacitor—it stores several (up to 20) coulombs of charge, and there is a large potential difference between it and the Earth. (The Earth acts like the second plate of an ordinary capacitor.) Air is normally a very good insulator, but if the p.d. across the air is large enough, the insulation breaks down because electrons become removed from some of the molecules. We say the air becomes **ionized**. When this happens, electricity flows between the cloud and the Earth in the form of a spark, or, more usually, several sparks in quick succession.

How can a thundercloud be charged to a high voltage? There is certainly no battery. Is there any way of charging things other than using a battery?

You may well have had the experience of pulling a nylon sweater over your head and hearing a crackling sound. This crackling is caused by small sparks. The nylon becomes charged with electricity as a result of rubbing against your hair. Rubbing an insulator like nylon (or various plastics) is the usual way of obtaining static electric charges. This phenomenon was known to the ancient Greeks, who rubbed natural resins, such as amber, to produce static electricity. The Greek word for amber is 'elektron'.

Notice that static charges can only build up on good insulators. Why?

The machine in Fig. 19.16 can produce quite high static voltages as a result of rubbing. The rubber belt becomes charged (by friction) as it passes round the bottom pulley. This charge is collected at the top (you will see how later) and flows onto the dome at the top.

The machine is called a **Van de Graaff generator**. Very large versions are used in atomic research to generate millions of volts. This high voltage can then be used to accelerate charged particles used for collision experiments which, as you will see in Chapter 21, form an important part of atomic physics. These machines have a high voltage generator to charge the belt; they do not rely on friction.

The p.d. produced by the small Van de Graaff generator in Fig. 19.16 can produce quite a respectable spark between the large sphere and the earthed sphere a few centimetres away—a miniature lightning flash.

This brings us back to the question of where a thundercloud gets its charge. Nobody is absolutely sure of the answer, but a thunderstorm tends to happen when a layer of cold air forms on top of a layer of warm air. The warm air tries to rise up through the cold air (Why?), producing strong currents of air within the thundercloud. These currents carry with them ice particles from the cloud, and it seems likely that friction between these ice particles could be responsible for the charge.

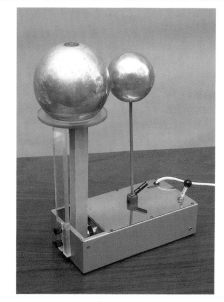

Fig. 19.16   A small Van de Graaff generator

Fig. 19.17   This large Van de Graaff generator is used by nuclear physicists to generate many millions of volts

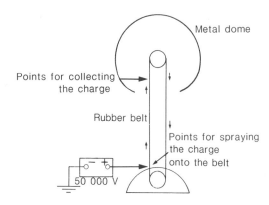

Metal dome

Points for collecting the charge

Rubber belt

Points for spraying the charge onto the belt

50 000 V

Fig. 19.18   A simplified diagram of a large Van de Graaff generator

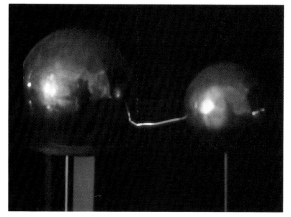

Fig. 19.19   A miniature lightning flash from a Van de Graaff generator

19.7 Electrostatic forces

You have probably tried rubbing your comb or a pen on your sleeve and then attracting tiny pieces of paper to it (Fig. 19.20), or rubbing a balloon and making it stick to the wall. Try rubbing your comb or pen and then hold it near (not touching) a thin trickle of water from a tap. What happens to the water?

As you know from the last section, the pen, comb or balloon is becoming charged, and you can see that electric charges can give rise to forces. Would you say that the forces are *small* or *large*?

Investigate these forces a little more as follows. Rub a strip of polythene with a woollen duster in order to charge it, and hang it in a wire stirrup as shown in the photograph. The stirrup should be suspended using a long nylon thread. Rub another piece of polythene and slowly bring it close to the first one. What happens?

Repeat this experiment with two strips of cellulose acetate. Does the same thing happen?

What happens if you repeat the experiment using one strip of polythene and one strip of cellulose acetate?

Fig. 19.20
Why is the paper attracted to Richard's pen?

You will have to do these experiments carefully; any forces are *very* small, and the charges tend to leak away after a time—more quickly on a damp day.

Why do you think charges will leak away more quickly on a damp day.

How can you explain the results of your experiments? It is the atoms on the surface of the polythene or cellulose acetate which are responsible for the static electricity. You have probably already learned that an atom has a positively charged centre (called the **nucleus**) surrounded by a cloud of negatively charged **electrons**. Scientists believe that when polythene is rubbed with wool, some of the electrons are 'scraped' off the atoms on the outside of the wool and move onto the polythene, making it *negatively charged*. (Remember that *only electrons* can move in this way.)

When cellulose acetate is rubbed, it becomes positively charged. Explain, in terms of electrons, what you think is happening this time.

In your experiments you found that the polythene and cellulose acetate, having opposite charges, attract each other. *Unlike charges attract.*

What do *like* charges do? Where have you come across a very similar rule before?

These rules about the forces between electric charges do not explain why a charged balloon sticks to a wall. A wall is not normally charged.

The answer to this problem is that the negatively charged balloon repels a few electrons from the surface of the wall as in Fig. 19.22, leaving the surface of the wall positively charged. The negatively charged balloon is then attracted to the positively charged surface of the wall.

We say a charge has been **induced** in the surface of the wall. The paper is attracted to the charged comb for the same reason.

Fig. 19.21
Investigating electrostatic forces

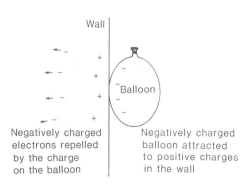

Negatively charged electrons repelled by the charge on the balloon

Negatively charged balloon attracted to positive charges in the wall

Fig. 19.22   How a charged balloon can stick to a wall

19.8 Electric fields

The area round a magnet where magnetic forces act is called a magnetic field. In a similar way, the region round a charged body where electric forces act is called an **electric field**. The shape of a magnetic field can be seen using iron filings. The shape of an electric field can be shown in a similar way, but not using iron filings; you can use semolina.

Fig. 19.23 shows how to do this. Half-fill a petri dish with olive oil and sprinkle some semolina on top. Fix the two metal plates in position and charge them from a high voltage power supply (or even from a Van de Graaff generator). Switch on the power. The semolina slowly moves to form the shape of the electric field.

Fig. 19.23   How to show the shape of an electric field using semolina floating on olive oil

Fig. 19.24 shows some examples of field patterns obtained in this way.

As with a magnetic field, the closer the field lines the stronger the field.

Any charged body in an electric field will try to move in the direction of a field line, as shown in Fig. 19.25. The 'direction' of an electric field is the direction in which a positively charged body will try to move, i.e. from the positive end of the field to the negative end of the field. (Remember that we also talk about the direction of a magnetic field. Which way does a magnetic field 'go'?)

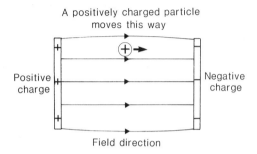

Fig. 19.25   The direction of an electric field

Look at Fig. 19.26. This apparatus is used to show up the effect of an electric field on a charged ball. There is an electric field between the plates. Why?

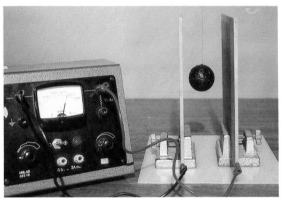

Fig. 19.26   What happens to this polystyrene ball when the electric field is switched on?

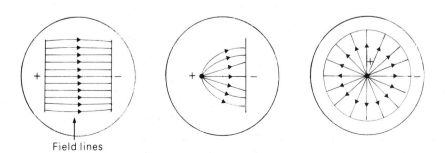

Fig. 19.24   The shapes of various electric fields

The polystyrene ball suspended between the plates is coated with graphite to make the surface conduct electricity well. If the ball is pushed to one side so that it touches one of the plates, the ball becomes charged. Why?

What happens if the ball is let go? Explain why this happens.

Take great care if you wish to do this experiment yourself instead of watching your teacher demonstrate it. The p.d. across the plates is 5 kV, although the power supply shown keeps the current down to a safe value so you cannot obtain a fatal shock.

Another way of showing up an electric field is to stick a small piece of aluminium leaf (not foil—leaf is much thinner and lighter) to a polythene strip as in the next photograph. In this photograph the leaf has touched the *left-hand* plate. Explain why it is now pointing to the right.

Fig. 19.27   Using a piece of aluminium leaf to show up an electric field

What would happen if the leaf touched the right-hand plate? Why?

Suppose, while the leaf is between the two plates and pointing right, the p.d. across the plates (and hence the electric field strength) is reduced. Will the position of the leaf change in any way?

19.9   The gold-leaf electroscope

Look at Figs 19.28 and 19.29. This instrument is used for detecting electric charges. It is called a **gold-leaf electroscope**. If the central column becomes charged, there will be an electric field

between it and the case of the electroscope. The fine leaf of gold stuck to the central column also becomes charged, so that it deflects for the same reason that the aluminium foil deflected in the last section. The bigger the charge on the gold-leaf electroscope, the bigger the potential difference between the central column and the outside, and the more the gold leaf deflects.

Fig. 19.28   A gold-leaf electroscope

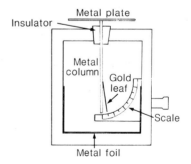

Fig. 19.29   The construction of a gold-leaf electroscope

Notice that the gold-leaf electroscope is really a kind of *voltmeter*, even though we are using it for detecting electric charge. Charge flows onto the electroscope and causes an electric field between the middle and the outside (like a capacitor); the gold leaf responds to this field.

Why do you think an ordinary moving-coil voltmeter would be useless for electrostatics' experiments?

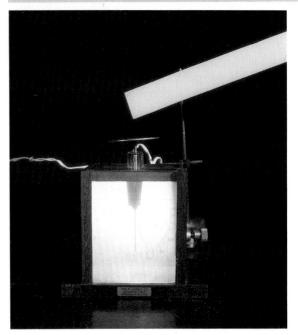

Fig. 19.30   A shadow of the gold leaf is being projected onto a
              ground glass screen

electroscope. Can you think why not? It is much better to charge the electroscope in the way shown in Fig. 19.32, which also shows what is happening to the electrons in the electroscope. The next diagram gives a similar explanation for a positively charged rod, such as cellulose acetate.

When trying to work out what happens in this sort of situation, it is important to remember that *only electrons can move*.

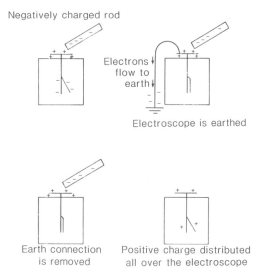

Fig. 19.32   Charging an electroscope by induction

The gold-leaf electroscope in the illustration has a ground glass screen at the back so that a shadow of the leaf can be thrown onto this screen. This makes it easier to see. It is being used in this way in the next photograph, where a charged polythene rod is being held near the electroscope.

But why is the leaf deflecting? The rod is not touching the electroscope. Fig. 19.31 explains what is happening. The leaf collapses as soon as the polythene rod is removed.

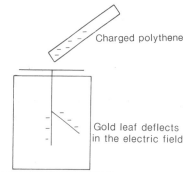

Fig. 19.31   Why the gold leaf in the last photograph is
              deflected

The electroscope can be permanently charged by *touching* it with the polythene. Try this for yourself. You will probably find that it is not very effective— not many electrons move from the polythene to the

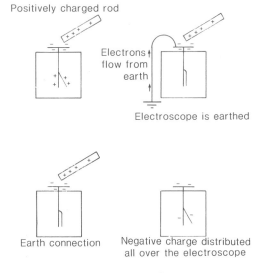

Fig. 19.33   Charging an electroscope by induction

Charging an electroscope in the way shown in Figs 19.32 and 19.33 is known as 'charging by induction'.

19.10 Two investigations

The two small investigations in this section will teach you two more important facts about electrostatic charges.

Charge a small metal can by putting it on top of a Van de Graaff generator and switching it on for a short time (Fig. 19.34). Switch off the generator. Is the charge distributed evenly all over the can? To find out, use a **proof plane** (a small metal disc on an insulated handle) to transfer some of the charge from the *outside* of the can to a gold-leaf electroscope (which should be some distance from the generator). Make a note of how much the leaf deflects. Discharge the electroscope (by touching the metal plate at the top with your finger) and then use the proof plane to transfer some charge from the *inside* of the can to the electroscope. Make a note of how much the leaf deflects this time. What conclusion do you come to from this experiment?

Fig. 19.34   Is the electric charge distributed evenly over this can?

Bearing in mind that like charges repel, can you suggest a reason for these results?

From the results of this experiment, where do you think all the charge on the dome of a Van de Graaff generator is? Can you think of a reason for bringing the charge up the belt to the *inside* of the dome?

For the second investigation, charge a pear-shaped conductor like the one in Fig. 19.35. Use a

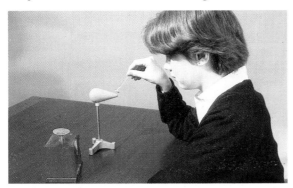

Fig. 19.35   Is electric charge distributed evenly over this conductor?

proof plane to transfer charge from different parts of the surface to a gold-leaf electroscope. Make a note of the deflection of the leaf each time, and remember to discharge the electroscope after each reading.

Does it seem to you as if the charge is evenly distributed over the surface of the conductor, or is there more charge in some parts than others.

19.11 Lightning conductors

Fig. 19.36   What is blowing this candle flame?

Look at Fig. 19.36. What is blowing the candle flame? The pin is connected to a Van de Graaff generator, so it is charged to a high voltage. One of the experiments in the last section will have shown you that the field round a sharp point is very strong. This strong field pulls some of the electrons off the atoms in the air nearby and the air becomes *ionized*. The electrons are attracted to the positively charged pin, and the positively charged ions are repelled from the point as a wind which blows the candle.

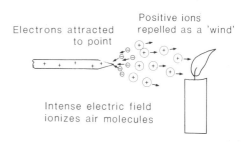

Fig. 19.37   How an electric wind is created

A similar effect occurs with a lightning conductor. A charged thundercloud causes the tip of the lightning conductor to become charged (with the opposite charge to that of the thundercloud). There is a strong electric field round the tip of the conductor and this ionizes some atoms in the air. The ions are attracted to the cloud and help to neutralize it, so it is less likely that there will be a flash of lightning. Even if there is a lightning flash, it is more likely to hit the lightning conductor than anywhere else, and the electric charge will be led safely to earth.

In Section 19.5 you learned that a Van de Graaff generator collects charge off the belt under the dome. It collects the charge using a row of points acting in the same way as the lightning conductor described above. Look for these points under the dome of a Van de Graaff generator at your school.

19.12 Uses of electrostatics

Electrostatics can be a nuisance and a danger on some occasions. For example, care has to be taken to avoid the build up of static electricity, with the consequent risk of sparks, in petrol tankers and oil refineries. Even the flow of liquids like oil, and many powders, can cause large electrostatic charges.

However, electrostatics is put to good use in some processes. Here are some examples.

Paint sprays

The car in the photograph is given an electric charge, and the drops of paint have the opposite charge so that they are attracted to the body of the car. Why do you think this method of spraying paint is used?

Electrostatic precipitators

A lot of ash is produced in a coal-fired power station. If nothing were done about it a lot of this ash would go up the chimney into the atmosphere. To prevent this, the smoke passes between charged plates. The ash is attracted to and sticks to these plates so that it does not pass into the atmosphere. Do you know what happens to the ash from a power station?

Xerox copiers

Many photocopiers work by putting an electrostatic charge on those parts of the paper where ink is required. Ink (in the form of a powder) then sticks to these charged parts of the paper but not the other parts. The ink is then 'set' into the paper by heating it.

Fig. 19.38   Electrostatics helps in spraying paint on car bodies

Fig. 19.39   Electrostatic precipitators being constructed at a power station

SUMMARY

Now that you have finished studying this chapter on static electricity, there are a number of things you should know and be able to do.

1 You should:
  a) know that a capacitor can store electric charge;
  b) know the definition of 1 farad of capacitance, and be able to use this definition in simple calculations;
  c) know the symbol for a capacitor;
  d) understand what a reservoir capacitor does;
  e) be able to explain (in terms of electrons) how an electric charge can be produced by rubbing;
  f) know that like charges repel and unlike charges attract;
  g) be able to charge something by induction;
  h) understand what is meant by an electric field;
  i) understand how a lightning conductor works;
  j) know some uses of electrostatics.

2 You should know what each of the following is, or what each does:
  Van de Graaff generator
  gold-leaf electroscope
  proof plane

3 These are some of the other words that have been used in this chapter. You should know what each word means:

  vivid   discharge
  version   deflect

FURTHER QUESTIONS

1 A large electrostatic charge can build up on a moving car, especially on a hot, dry day.
  a) How do you think the car becomes charged?
  b) Explain why it does not become charged on a wet day.
  c) What can be done to reduce the build-up of charge?

2 Fig. 19.40 shows two light, conducting balls suspended by nylon threads. They are at rest with the threads making an equal angle to the vertical as shown. In trying to explain this, some pupils made the following statements. Comment on these statements and say whether each one is true, untrue, or could be true.
  a) The balls have equal negative charge.
  b) The balls have equal positive charge.
  c) The balls do not necessarily have equal charges.
  d) One ball must be uncharged.
  e) The balls must have the same mass.

3 Fig. 19.41 shows a metal can on top of a gold-leaf electroscope. The inside of the can contains a polythene rod and some wool as shown. When the polythene rod is rotated backwards and forwards in the wool, there is no effect on the gold-leaf electroscope. When the rod is rotated and then removed from the wool, the leaf of the electroscope diverges. When the rod is replaced, the leaf collapses again. Explain these effects.

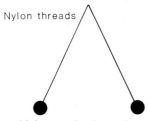

Nylon threads

Light, conducting spheres

Fig. 19.40

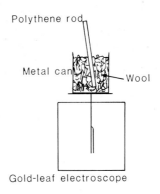

Polythene rod

Metal can — Wool

Gold-leaf electroscope

Fig. 19.41

20

RADIOACTIVITY

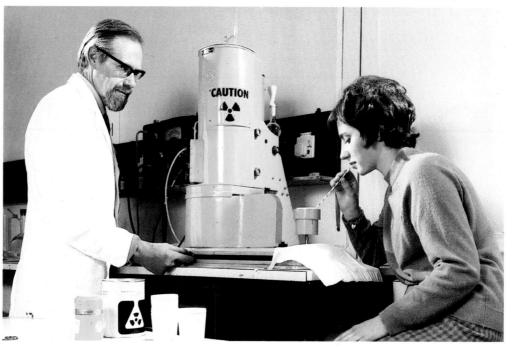

Fig. 20.1   This patient's thyroid gland is being investigated with the aid of a radioactive 'tracer'

20.1   Introduction

The drink that the hospital patient is having in Fig. 20.1 contains an iodine compound. The iodine will concentrate in the thyroid gland, and the amount of iodine which collects in the gland will later be measured in order to check how well the gland is working.

But how will the concentration of iodine in the gland be measured? Certainly not by any method which involves surgery. The iodine is a special sort of iodine which gives off a form of radiation. This radiation can be detected (Fig. 20.2) and so the concentration of iodine can be measured. The iodine which gives off this radiation is **radioactive**.

What is radioactivity? In what ways are radioactive atoms different from normal atoms? Are there any other radioactive atoms apart from iodine ones? How do radiation detectors work? Does the iodine change in any way when it gives out radiation? Could the radiation do the patient any harm? By the time you have finished this chapter you should be able to answer all these questions.

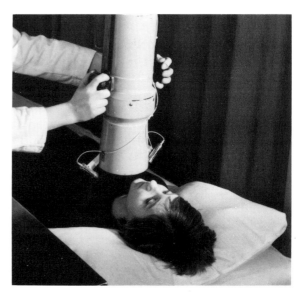

Fig. 20.2   This machine detects the radiation from the radioactive tracer that has collected in the thyroid gland

The fact that some substances give out natural radiation was discovered by accident in 1896 by a French scientist called Becquerel. He found that some photographic plates, properly wrapped in black paper, had become fogged. The plates had been stored next to some uranium ore, and Becquerel traced the cause of the fogging to this uranium ore.

It was soon discovered that other substances gave off this radiation which would fog photographic plates. In some cases the radiation was much more powerful than that from uranium. In 1898 Madame Curie, working in Paris, extracted the radioactive elements radium and polonium (the latter named after her native country of Poland) from an ore called pitchblende.

To attempt to follow the early work on radioactivity in any depth would be very confusing—the early workers were confused themselves, and it took a number of years before a clear picture of radioactivity began to emerge.

Fig. 20.3   Why does radium make this gold-leaf electroscope discharge?

20.2   You should know . . .

You will understand this chapter best if you:
1 understand what happens to an atom when it is ionized;
2 understand Sections 19.7 and 19.8 about electric fields and forces;
3 know what happens to a wire carrying an electric current through a magnetic field.

20.3   Detecting radiation

The radiation from a radioactive substance cannot be seen, so how do we know it is there; in other words, how do we detect it? Becquerel discovered one way, of course, in his accidental discovery of radioactivity, but much better and more useful methods now exist.

If a radioactive substance, such as radium, is held near a charged gold-leaf electroscope as in Fig. 20.3, the electroscope discharges. Why should this happen? The same thing happens if you use a lighted match instead of a radioactive substance. The energy from the match flame is sufficient to *ionize* some of the air molecules. If the electroscope is positively charged, electrons which have been stripped off the air molecules as they were ionized flow onto the electroscope and neutralize the positive charge.

Fig. 20.4   A lighted match will also discharge a gold-leaf electroscope

Write down what you think happens if the electroscope is negatively charged.

The radium discharges the electroscope for the same reason. The radiation from radium has sufficient energy to ionize air molecules.

This is not a very *practical* way of detecting radiation, but the important point to learn from this simple experiment is that radiation from radioactive substances has sufficient energy to ionize atoms and molecules. All the equipment you will meet in this chapter for detecting or measuring radiation relies on this fact—radiation causes ionization.

227

20.4 Safety

Since the radiation from radioactive materials has enough energy to ionize air molecules, for example, it also has sufficient energy to change the molecules in your body. The radiation can ionize your molecules, or change the structure of the cells in some way. The results can include radiation 'burns' and tumours in the cells.

This is one of the hazards of radioactivity. Although the radioactive sources used in schools are not particularly strong, they must still be handled properly. This is why the radium in Fig. 20.3 is being handled with tongs, and not picked up directly, so that it is kept a safe distance from the fingers. There are rules prohibiting young people under 16 handling most radioactive substances. This is why you will find your teacher demonstrating most experiments in this chapter.

In places where stronger sources are used, operators must be shielded from the sources by, for example, lead or concrete. Strong sources have to be handled by remote control.

20.5 From spark counter to Geiger-Müller tube

The apparatus being used in Fig. 20.5 consists of a fine wire stretched below a metal grid, as shown in the diagram. The p.d. between the wire and the grid is about 4 kV, the wire being positive. You can see in the photograph that the radium is causing sparks to fly between the wire and the grid. The sparks stop as soon as the radium is taken away.

Will a lighted match also cause sparks in this 'spark counter'?

The simple explanation for the sparks is that the radiation is ionizing the air so that it conducts electricity. However, it is really a little more complicated than that. If an air molecule in the space between the grid and the wire becomes ionized, the electron, having a very small mass, will rapidly accelerate towards the positive wire. On its way it will most likely collide with another air molecule. The electron will probably have gained sufficient kinetic energy to be able to ionize this molecule, releasing a second electron to accelerate towards the wire. This process repeats itself (Fig. 20.7) many times, thus causing an *avalanche* of electrons (which appears as a spark).

Suppose you use a spark counter with a different radioactive source? Common sources in schools include isotopes of cobalt, americium, plutonium

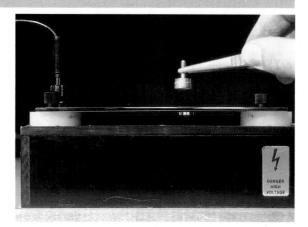
Fig. 20.5   A spark counter in use

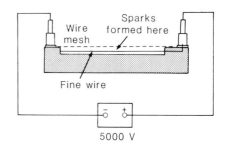
Fig. 20.6   The construction of a spark counter

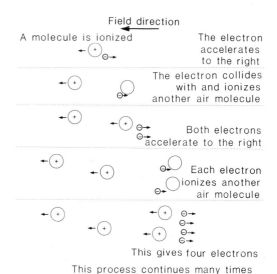
Fig. 20.7   An avalanche of electrons

and strontium as well as the radium used in Fig. 20.5. Do you obtain the same results with these other sources? If there are any differences, can you suggest possible reasons for them?

You will probably agree that a spark counter is not a very *convenient* device for detecting radio-activity. It lacks sensitivity (What does that mean?) and needs a high voltage to operate. However, the idea behind the spark counter can be developed into a more practical device, the **Geiger-Müller tube**.

A diagram of a Geiger-Müller tube appears in Fig. 20.8. If you look at this diagram you can see that there is a fine wire down the middle of the tube (this corresponds to the fine wire in the spark counter). The cylinder surrounding this wire is metal (this corresponds to the grid in the spark counter). The thin mica window at the end is to allow the radiation in. The inside of the tube is at a low pressure, most of the air having been pumped out, so the mica is needed to seal the end. The low pressure makes the tube more sensitive than a spark counter, and also enables it to be driven off a lower voltage—the p.d. between the wire and the case is usually about 400 V, the wire again being positive.

As with the spark counter, radiation causes pulses of electricity ('sparks') to cross the gap between the case and the wire. You cannot hear these sparks. The pulses can be fed to an amplifier and loud-speaker so that you can hear them, or, more usefully, they can be fed to either a **scaler**, which counts the total number of pulses, or to a **ratemeter**, which measures the number of pulses per second. The instrument illustrated in Fig. 20.9 is both a scaler and ratemeter combined. It has its own high voltage supply for the Geiger-Müller tube. In the photograph it is displaying the total number of pulses it has received since it started counting.

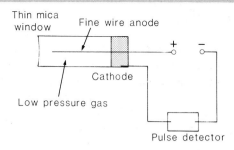

Fig. 20.8   The construction of a Geiger-Müller tube

Fig. 20.9   A Geiger-Müller tube connected to a scaler

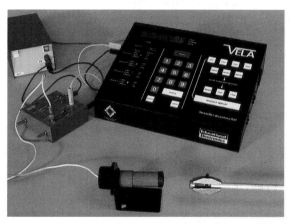

Fig. 20.10   Detecting the radiation from plutonium. The VELA is acting as a ratemeter

20.6   Some properties of radiation

It was by no means an easy task for scientists at the turn of the century to find out exactly what the radiation from radioactive substances is. To describe all the experiments and false ideas would be too confusing. In this section you will look at experiments which tell you about two important properties of radiation which provided scientists with some of the evidence from which they were able to deduce the identity of the radiation.

Penetration

Radiation from radioactive substances will go through (penetrate) some substances. Does it go through all substances equally well? Does it make any difference what substance is producing the radiation?

To answer these questions, plutonium or americium should be supported near the window of a Geiger-Müller tube as in Fig. 20.10. The Geiger-Müller tube should preferably be connected to a ratemeter, although a scaler or amplifier and loud-speaker will do. In Fig. 20.10, the VELA has been instructed to act as a ratemeter. After the power has been switched on and everything has settled down, make a note of the count rate.

229

Does it make any difference to the count rate if a sheet of paper is put between the source and the detector? If not, suppose you use something thicker, like a sheet of cardboard? Does aluminium foil make any difference? A thicker piece of aluminium? A piece of lead? Does it make any difference if the source is closer to the Geiger-Müller tube, or moved away from the tube?

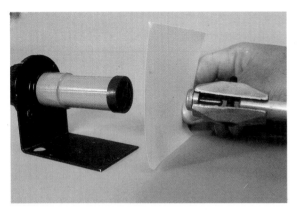

Fig. 20.11   Will the radiation go through this paper?

If a different radioactive source is used and the experiment repeated, are the results the same? If possible, compare the results of using plutonium or americium with strontium and cobalt. Then try radium (which gives more complicated results).

Is there any evidence for more than one kind of radiation? If so, how many kinds?

The effect of a magnetic field

For this investigation a strontium source should be supported about 15 cm from a Geiger-Müller tube as in Fig. 20.12.

Make a note of the count rate when the power to the Geiger-Müller tube has been switched on.

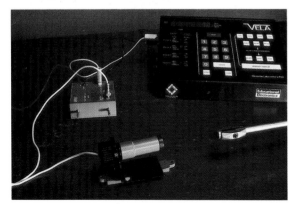

Fig. 20.12   The radioactive source for this investigation is strontium

A powerful magnet should now be placed between the source and the tube (Fig. 20.13) so that the radiation passes between the poles of the magnet. (The height of the source and tube may need adjusting.) Does the count rate stay the same? If not, what has happened to the radiation? Has it been deflected in a different direction (Fig. 20.14) or is it just lost? Does it make any difference if the magnet is turned over?

Fig. 20.13   Does the magnet have any effect on the radiation?

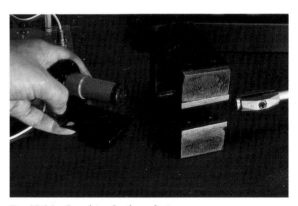

Fig. 20.14   Searching for the radiation

Can you remember seeing a magnetic field have the same effect on anything else?

The experiment should now be repeated with cobalt. Does the same thing happen?

It is not possible to do this experiment with americium or plutonium unless the whole experiment is done in a vacuum. (Can you think why not?) Scientists have found that radiation from these two substances is bent *very slightly* by a magnetic field, and it is bent in the opposite direction to the radiation from strontium. Can you come to any conclusion from these observations?

20.7 What are the different radiations from radioactive substances?

The experiments described in the last section are just two examples of experiments that led scientists to the conclusion that there are *three different kinds* of radiation from radioactive substances. They are known as **alpha** (α), **beta** (β) and **gamma** (γ) radiation.

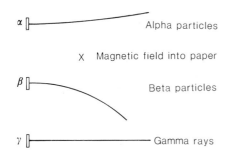

Fig. 20.15   How some different radiations behave in a magnetic field

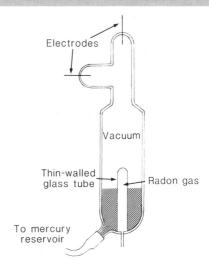

Fig. 20.16   The Rutherford–Royds experiment

Alpha radiation

This is the radiation from plutonium and americium. It is stopped by a few centimetres of air or a sheet of paper, and deflected a little by both a strong magnetic field and a strong electric field, in such a direction as to suggest it is a *particle* with a *positive charge*.

Measurements of **charge-to-mass ratio** (the number of coulombs for every kilogram) at the beginning of the twentieth century suggested that an alpha particle might be a helium atom without its two electrons, i.e. a helium ion with a double positive charge. This was confirmed in 1909 by Rutherford and Royds using the apparatus illustrated in Fig. 20.16.

In their apparatus, Rutherford and Royds compressed some radioactive gas called radon in the thin-walled tube as shown. Radon emits alpha particles, which escaped through the thin wall into the evacuated space outside. The walls enclosing the evacuated space were too thick for the alpha particles to go through, so the alpha particles were collected in the evacuated space.

After about a week, the mercury level was raised, so compressing any gas which had collected in this space to the top, where there was a pair of electrodes. An electric current was passed through the gas and they observed the characteristic spectrum of helium. They concluded that an alpha particle

is an ionized helium atom; each alpha particle collected two electrons (Where from?) and became a helium atom.

What sort of equipment do you think Rutherford and Royds used to observe the helium spectrum caused by the spark?

Beta particles

This is the radiation from strontium. As you observed in the last section, it is moderately penetrating, needing a few millimetres of aluminium to stop it totally. It is deflected by both magnetic and electric fields in such a direction as to suggest it is a negatively charged particle.

Measurements showed that it has a charge-to-mass ratio exactly equal to that for electrons, and it was concluded that beta radiation consists of electrons travelling at very high speeds—up to 90% of the speed of light.

Gamma radiation

This is the radiation which comes from cobalt. It is very penetrating. The fact that it is not deflected by a magnetic or electric field indicates that it consists of either uncharged particles, or waves. In 1914 gamma rays were shown to give interference and diffraction effects, thus indicating they are waves. They are part of the electromagnetic spectrum with a very short wavelength, as you will already know if you have read the section on the electromagnetic spectrum (Section 12.9).

To summarize this information about the different kinds of radiation, copy out and complete the table below.

Name of radiation	Typical source	What stops radiation?	Effect of magnetic field	Identity
Alpha				
Beta				
Gamma				

20.8 Vapour trails

Fig. 20.17 Vapour trails in the sky

You must have seen a sight similar to that shown in Fig. 20.17—vapour trails in the sky formed by high flying aircraft. How do they get there? Why are they not always formed?

There is always some water vapour in the atmosphere as a result of evaporation from rivers and seas. The amount of water vapour that the air can hold depends on the temperature—the colder it is, the less water vapour can be held in the atmosphere. If the atmosphere is holding as much water vapour as it can at a particular temperature, it is said to be *saturated*. If it gets colder, some of this water vapour will condense, clouds will form, and it may well start raining.

However, it is not easy for water to condense unless it has something on which it can condense. Dew is formed on the ground because, as night falls, the air temperature decreases and the atmosphere may become saturated with water, so water condenses on the ground. High in the atmosphere there may be nothing on which the water vapour can condense, and the atmosphere may become **supersaturated**, that is, hold more water vapour than it normally would. As soon as there is something onto which the water vapour can condense, it will quickly do so.

The water vapour can condense onto the particles in a high flying aeroplane's exhaust, so if an aeroplane flies through atmosphere that is supersaturated, a vapour trail will form behind it, showing where the aircraft has been. If you look carefully at a vapour trail next time you see one being made, you will see that once it has been started, it becomes wider and wider as more water vapour condenses; it just needed something to start it off.

20.9 The diffusion cloud chamber

What have vapour trails to do with radioactivity? Water vapour will easily condense on ionized molecules. Since the radiation from radioactive substances leaves a trail of ionized molecules behind it, perhaps we could follow the path of the radiation by making little vapour trails in a similar way to those made by aircraft.

To do this, we must make some air supersaturated with water vapour. This can be done in a piece of apparatus called a **cloud chamber**. An example is illustrated in Fig. 20.18; it is called a **diffusion cloud chamber**. To use it, the bottom of the chamber must be made very cold. To do this, unscrew the base, remove the foam plastic and place some solid carbon dioxide (dry ice) underneath the floor of the chamber, then replace the foam plastic and lid.

Take care not to touch the dry ice—you could get severe frostbite.

Fig. 20.18 A diffusion cloud chamber

Make sure that the special weak radioactive source for use with this cloud chamber is inserted in the side of the chamber; this source gives out alpha particles.

Soak the felt ring round the top of the chamber with a meths/water mixture by means of a pipette, and replace the perspex lid. This mixture will soon vaporize and the air at the top of the chamber will become saturated. This saturated vapour drifts down to the cooler air at the bottom of the chamber, where it becomes supersaturated. Any radiation coming from the source will leave trails of ionized air, giving something for the supersaturated vapour to condense on.

The cloud chamber should be illuminated from the side as in Fig. 20.19, and allowed to settle down. After a short while vapour trails similar to those in Fig. 20.23 should appear.

So that the inside of the chamber does not become a confused muddle of vapour, the ions which are formed should be cleared quickly, so that you only see the vapour trails on the recently formed ions. This can easily be done by rubbing the lid of the chamber with a cloth every so often. This creates an electric field which pushes the ions out of the way.

Look carefully at the vapour trails. Remember that these vapour trails mark where the alpha particles have been. You are not looking at the radiation itself.

20.10   Cloud chamber tracks

The tracks that you have looked at, like the ones shown in Fig. 20.23, are made by alpha particles. They are bright and not very long—usually less than the diameter of the cloud chamber. They are bright because alpha particles cause a lot of ionization per centimetre of track length. For this reason

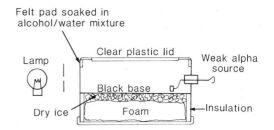

Fig. 20.19   The construction of a diffusion cloud chamber

Fig. 20.20   Putting dry ice in a diffusion cloud chamber

Fig. 20.21   Inserting a radioactive source into the side of the chamber

Fig. 20.22   Soaking the felt ring with a meths/water mixture

Fig. 20.23   Tracks in a cloud chamber

the alpha particles lose their energy in a short distance, which is why the tracks are not very long. It is also the reason why alpha particles are able to form sparks with the spark counter (beta and gamma radiations, you remember, do not). Alpha particles from any particular source usually have the same energy, so all the tracks will be of the same length.

Were all the tracks in your cloud chamber of the same length?

The alpha tracks are straight for most of their length, but towards the end of the track, when they are moving relatively slowly, the alpha tracks often have small changes of direction, as shown in Fig. 20.24. What do you think causes these changes in direction of the alpha particle?

Tracks from beta particles are more difficult to see. Fig. 20.25 shows a typical example of the thin, straight track formed by a fast beta particle. Do you think the beta particle causes more or less ionization per centimetre than the alpha particle? Will a beta particle lose all its energy over a longer distance or a shorter distance than an alpha particle? Does this agree with the fact that beta particles are more penetrating than alpha particles?

The next photograph shows the track of a beta particle while in a magnetic field. In what way does this confirm the results of the experiment with beta radiation in a magnetic field described in Section 20.6? In which direction was the magnetic field when this photograph was taken?

Fig. 20.26   Beta tracks in a magnetic field

Fig. 20.27 shows the effect of a gamma ray in a cloud chamber. The tracks you can see are caused by electrons ejected from air molecules by the gamma ray. They are ejected with so much energy that they can cause quite a lot of **secondary ionization** of air molecules.

Fig. 20.27   Cloud chamber tracks caused by gamma rays

20.11   Collisions

The inside of an atom still holds many secrets, and one of the most useful tools to a physicist trying to unlock those secrets is the **bubble chamber**. (Try to find out how a bubble chamber works; in some ways it is similar to a cloud chamber.) These physicists fire atomic 'missiles' (e.g. high-speed alpha particles, or high-speed electrons) at atoms

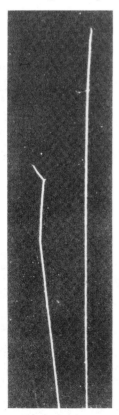

Fig. 20.24
A magnified view of the end of a track caused by an alpha particle

Fig. 20.25
A cloud chamber track caused by a beta particle

in bubble chambers, and look at the results of any collisions which take place. Much of the money spent on this kind of research is devoted to building ever more powerful particle accelerators so that the particles can penetrate ever further into the atoms.

In a cloud chamber alpha particles bump into air molecules and ionize them. Nearly all these collisions are 'glancing blows', just 'chipping' an electron off the outside of the atom. Very occasionally there is an almost 'head-on' collision with an atom. Such collisions gave scientists, like P. M. S. Blackett working in the 1920s, valuable information about alpha particles.

Fig. 20.28 shows three well known examples of Blackett's early photographs of collisions between alpha particles and atoms. In Fig. 20.28(a), the cloud chamber was full of hydrogen; an alpha particle had a nearly head-on collision with a hydrogen atom and gave it so much energy that it was able to cause ionization of its own, so you can see the track of the recoiling hydrogen atom. The important thing, however, is the angle between the two tracks. In this case it is an acute angle (less than 90°). This indicates that the alpha particle is more massive than the hydrogen atom.

Fig. 20.28(b) shows a similar collision with a nitrogen atom. In this case the angle between the tracks after the collision is obtuse. It is possible to show that this means that the alpha particle is less massive than a nitrogen atom.

The most important photograph is Fig. 20.28(c). Helium was in the cloud chamber, and the angle here is a right angle. This indicates that an alpha particle has the same mass as a helium atom. Why do you think this was an extremely important photograph? What has it to do with the Rutherford–Royds experiment?

20.12 Uses of radioactivity

So far in this chapter we have largely concentrated on *explaining* radioactivity, but radioactive substances have many practical uses. In this section you are going to have a quick look at just a few of these (there are so many that a whole book could be written on this subject alone).

At the start of this chapter you saw one medical use of radioactivity, where iodine was being used as a **tracer**. The whereabouts of the radioactive iodine inside the body could be traced from the outside. Radioactive sodium in common salt is used to trace the passage of blood round the body to look for blocked blood vessels. Radioactive phosphorus (in phosphate fertilizer) is used to follow the take up of fertilizers in plants. Radioactive substances are added to the silt on sea or river beds so that the drifting of the silt can be monitored. Why might engineers want to do this?

Fig. 20.29 Checking the build up of silt on the sea bed with the aid of a radioactive tracer

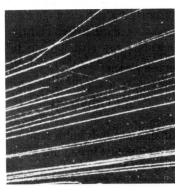

Fig. 20.28(a) A collision between an alpha particle and a hydrogen nucleus

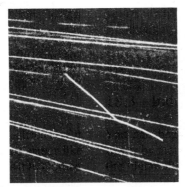

Fig. 20.28(b) A collision between an alpha particle and a nitrogen nucleus

Fig. 20.28(c) A collision between an alpha particle and a helium nucleus

Thickness of paper while it is being manufactured is checked using a beta source, as shown in Fig. 20.30. If the paper is too thin, the count rate at the detector will be too high; if the paper is too thick the count rate will be too low. The rollers which control the thickness of the paper are usually adjusted automatically as a result of the count rate at the detector.

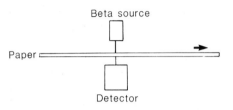

Fig. 20.30 How a beta source can be used to monitor the thickness of a sheet of paper as it is being made

The same technique is used in the manufacture of other sheet materials, e.g. rubber. A similar technique is used to monitor the thickness of a layer of paint, or check that sealed cans are full (Fig. 20.31).

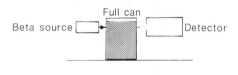

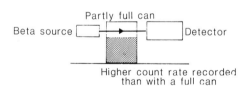

Fig. 20.31 Checking that sealed cans are full

Leaks in underground pipes are detected using radioactivity. A material emitting beta particles is dissolved in the liquid flowing along a pipe at the start of the pipe. The beta particles cannot get through the pipe, but any radioactive substance leaking from a crack can be detected at ground level with a suitable detector.

Gamma rays, usually from cobalt, are often used instead of X-rays to inspect, for example, welded seams in pipes. Gamma rays are easier to use than X-rays in that a gamma source is easily portable, so it can be used 'on site', whereas X-rays require high voltages and bulky equipment.

Gamma rays from cobalt are often used in hospitals for the treatment of cancers. The gamma rays kill the cancer cells, but care obviously has to be taken to minimize damage to sound cells (Fig. 20.32).

Gamma rays will kill bacteria and are often used for sterilization of, for example, medical dressings and hypodermic syringes. This can be done after they have been packed (since gamma rays will easily pass through the packing), so eliminating any possibility of their becoming contaminated between being sterilized and being packed. This technique is also valuable when sterilization by hot water or steam would damage the thing being sterilized.

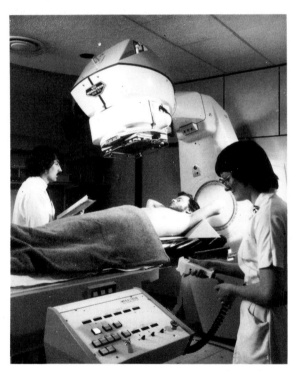

Fig. 20.32 Treating cancer cells with gamma radiation from radioactive cobalt

20.13 Radioactive decay

Is it not a little dangerous using radioactive tracers in people? You have read of some of the precautions to be taken when using radioactive substances to minimize the risk of their causing damage to people, yet now we see examples of people swallowing or being injected with radioactive substances! In fact, the quantity of substance used is very small, so the risk is not great. Also, all radioactive substances become less active as time goes on—the activity **decays**.

One possible substance that can be used in the laboratory to investigate this decay is a gas called radon. If some thorium hydroxide powder is kept in a polythene bottle, radon gradually collects in the bottle, and some can be obtained for this investigation by simply squeezing the bottle. Radon emits alpha particles.

The aim of this experiment is to draw a graph of the activity of the radon against time. A Geiger-Müller tube is used to detect the alpha particles.

There are a number of possible ways of obtaining data for a graph.

1 Connect the Geiger-Müller tube to a ratemeter and record the count rate every 15 seconds. Draw a graph of count rate against time.
2 Connect the Geiger-Müller tube to a scaler; note the scaler reading every 15 seconds, and quickly reset the scaler for the next 15 seconds. Draw a graph of the counts in 15 seconds against time. Alternatively, if resetting the scaler takes too long, it can be left running, and the total counts from the start of the experiment recorded every 15 seconds. Each reading must then have the previous one subtracted from it to find the counts in each 15 second period before plotting a graph as described above.
3 Connect the output from the Geiger-Müller tube to a VELA, which can be instructed to plot a graph on an oscilloscope of the number of counts in 15 seconds against time.

Fig. 20.33 shows this experiment being performed using a VELA to record the results. The experiment was carried out as follows.

1 The Geiger-Müller tube was placed in the top of the flask and the background count rate was measured over a period of at least a minute. The background count per 15 seconds was calculated. The background count had to be subtracted from any reading using radon, since the Geiger-Müller tube always picked up background radiation as well.

Why do you think the background was measured for a period of a minute or more, rather than for just 15 seconds?

2 A 'radon generator' was connected to the flask as shown, the screw clips were undone and the generator was squeezed gently a couple of times to transfer some radon to the flask. The screw clips were re-tightened.

3 Readings of counts every 15 seconds were taken by the VELA for about 5 minutes. The VELA drew a graph of the results on the oscilloscope as they were recorded. If other means are used to record the results then a graph of counts per 15 seconds against time should be plotted at the end of the experiment, making due allowance for background.

The Geiger-Müller tube should be left sealing the top of the flask for at least an hour.

Fig. 20.34 is the chart recorder output of the results recorded by the VELA. Notice how the activity decreases with time, but the graph is not a straight line; the activity decreases a lot to start with, and more gradually later. Because of the shape of the graph, you will agree that it is not really possible to say how long it took all the radon to decay, so we refer to the **half-life** of the radon, that is, the time taken for the activity to halve.

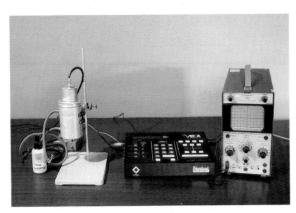

Fig. 20.33 Investigating the decay of radon

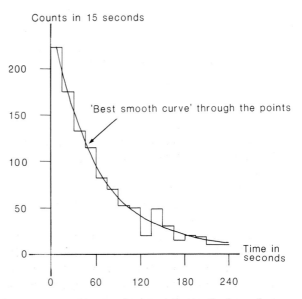

Fig. 20.34 A chart recorder output showing the decay of radon

Notice the somewhat irregular nature of the graph. Radioactive decay is a random process, and the graph shows the typical random fluctuations you can expect from any radioactivity experiment.

Look again at Fig. 20.34. You can see that, at the start, there were 220 counts in the first 15 seconds. A count rate of half that figure, i.e. 110, occurred after 45 seconds, so the half-life of radon, according to this experiment, is 45 seconds. After another 45 seconds, half of what is left has decayed, and the count rate is one-quarter of that at the start, that is, 55 counts in 15 seconds. After another 45 seconds there is only one-eighth of the radon left. Notice that it has *not* all gone after two half-lives (i.e. 90 seconds). In theory, it never all goes; it just goes on halving every 45 seconds. In practice, the quantity of radon becomes too small to detect after some time. The graph shows an **exponential** shape; we say the decay is exponential.

Every radioactive substance has its own half-life. Some half-lives are very long indeed, uranium-238, for example, has a half-life of 4.5×10^9 years; others are very short, polonium-214 has a half-life of 1.6×10^{-4} seconds. Most half-lives lie somewhere between these two extremes.

SUMMARY

Now that you have finished studying this chapter on radioactivity, there are a number of things you should know and be able to do.

1 You should:
   a) know that the radiation from radioactive substances can be detected because it has sufficient energy to ionize atoms;
   b) know what safety precautions should be taken when handling radioactive substances;
   c) know how a Geiger-Müller tube is constructed, and be able to explain how it works;
   d) know the identity of alpha, beta and gamma radiation;
   e) know the penetrating ability of the three kinds of radiation;
   f) know the behaviour of the three kinds of radiation in a magnetic field;
   g) understand the principle of Rutherford and Royds' experiment to identify alpha particles;
   h) know how a diffusion cloud chamber is constructed, and be able to explain how it works;
   i) understand how vapour trails form;
   j) understand the evidence from cloud chamber collisions for the identity of alpha particles;
   k) know some examples of the uses of radioactive tracers;
   l) know what is meant by half-life, and be able to measure the half-life of a substance from a suitable graph;
   m) be able to describe one experiment to determine the half-life of a radioactive substance.

2 You should know what each of the following is, or what each does:

spark counter	mica
avalanche	ratemeter
(of electrons)	exponential shape
scaler	(of graph)
dry ice	

3 These are some of the other words that have been used in this chapter. You should know what each word means:

ionize	prohibit
tumour	penetrate
saturated	supersaturated
sterilize	eliminate

FURTHER QUESTIONS

1 When a radioactive source which emits alpha particles is held close to a spark counter, a lot of sparks are observed. As the source is moved away from the spark counter, at a certain distance the sparks suddenly stop altogether. How do you explain this?

2 Alpha tracks in a cloud chamber are usually straight, but each alpha particle will collide with as many as 100 000 atoms along its path.
   a) What does this suggest to you about the atoms of gas in the cloud chamber?
   b) Why do the alpha particles eventually stop?

3 a) The half-life of radium is 1620 years. Explain what this means.
   b) If a sample of radium has a mass of 2 g today, what mass of radium will there be in 3240 years time?
   c) How long will it be before there is $\frac{1}{8}$ g of radium left?

4 a) Explain why a radioactive substance which emits gamma rays is more dangerous, and requires more careful handling, than one which emits alpha particles.
   b) However, a radioactive substance which emits alpha particles which is accidentally swallowed is much more dangerous than one which emits gamma rays. Why?

5 Hospital supplies such as hypodermic needles are often packed in airtight plastic bags and then subjected to radiation from a radioactive source in order to sterilize them after the bag is sealed.
   a) What kind of radiation would be used?
   b) What are the advantages of this method of sterilization over sterilizing in boiling water before packing?
   c) What precautions would have to be taken while sterilizing things in this way?

6 Two metal plates are connected to the positive and negative terminals of a power supply as shown in Fig. 20.36. When a lighted candle is placed in the gap between the plates, the flame and hot air spread out as shown. Suggest why.

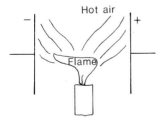

Fig. 20.35

7 Fig. 20.37 shows a radioactive source R contained at the bottom of a narrow lead cylinder. The source emits radiation which passes between the poles of a strong magnet. Two similar Geiger-Müller tubes, connected to separate counters, are placed at P and Q as shown. The two tubes point towards the gap between the poles of the magnet.

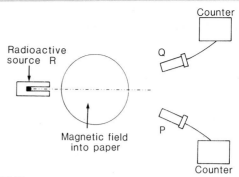

Fig. 20.36

The number of counts per minute, indicated by each counter, is recorded at 10-minute intervals. The results are tabulated below.

	Counts per minute at:	
Time in minutes	Counter P	Counter Q
0	585	42
10	405	39
20	290	40
30	205	38
40	155	43
50	120	37
60	95	41

a) Explain why the count rate is so much higher at P than it is at Q.
b) Plot two graphs on the same axes to show how the count rate (y-axis) changes with time (x-axis).
c) Suggest a reason for the shape of each graph. What radiation is being detected in each case?
d) Estimate the count rate at counter P after 1 hour more.
e) Estimate the count rate at counter Q after 1 hour more.
f) What experimental results would you expect to get if the magnetic field were:
   i) reversed;
   ii) removed altogether?
g) Estimate the half-life of the radioactive source.
(Nuffield O-Level Physics)

21

INSIDE THE ATOM

Fig. 21.1   CERN—the atomic research laboratories at Geneva

21.1   Introduction

By the end of the nineteenth century, nearly 100 elements had been discovered. Scientists believed that these elements consisted of atoms, and that the atoms of one element were totally different from atoms of another element. They thought atoms could not be cut up. Then, at the beginning of the twentieth century, a scientist called Thomson discovered that all atoms contained an even smaller particle, the electron. Electrons were absolutely identical no matter from which atom they had come. This discovery was the start of **atomic physics**.

Soon people like Thomson and Rutherford had built up a simple picture of what an atom might

look like inside. (You will look at some of the details in this chapter.) But the more closely scientists have examined the atom, the more complicated it has become. More and more subatomic particles have been discovered. The main way of unlocking the secrets of the atom is to smash it with very high energy 'missiles' such as alpha particles, and have a look at what comes out of the collision, in much the same way that Wilson looked at collisions between alpha particles and atoms in his cloud chamber.

The photograph on this page shows the nuclear research laboratories at Geneva, in Switzerland, which has facilities for accelerating atomic 'missiles' to very high energies before smashing them into atoms. Building such accelerators is very expensive, and requires a lot of space. The laboratories here are supported by several European governments so that the costs can be shared as much as possible.

21.2 You should know . . .

In order to understand this chapter properly, you should understand and know about:
1 electrical energy;
2 electrostatic fields and forces;
3 the catapult effect;
4 Chapter 20 (on radioactivity).

21.3 The discovery of the electron

We have used the idea of electrons extensively in this book to explain electric currents and static electricity. But how are scientists so sure that electrons exist? They are, after all, far too small to be seen.

The evidence for electrons built up gradually over many years, with many different scientists making contributions. An important experiment in this field was performed by J. J. Thomson in 1897. He was working in the Cavendish Laboratory at Cambridge and was investigating **cathode rays** produced by a **discharge tube** like the one illustrated in Fig. 21.2. The tube had a low pressure gas inside it. When a high voltage was connected between the anode and cathode, a current flowed and the tube glowed with a colour characteristic of the gas in the tube. (This is the principle behind the modern coloured lights that are used, for example, in advertising signs.)

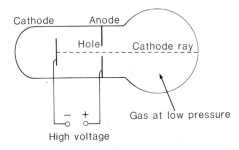

Fig. 21.2 Thomson's discharge tube

Thomson's electrons were produced by *collisions*. A chance collision might be enough to ionize one atom in the tube. The electron would accelerate towards the anode, and, on the way, would collide with further atoms producing more electrons. These electrons would, in their turn, collide with atoms to produce further electrons. The result was a continuous stream of electrons heading for the anode.

What do you think happened to the positive ions produced by these collisions?

Thomson knew nothing of the above explanation. All he knew was that these mysterious 'cathode rays' were produced. He thought these rays could consist of tiny particles. He was subsequently able to measure the charge-to-mass ratio for these rays. The fact that he was able to do so suggested that these rays did have a mass, and lent support to the idea that the rays consisted of a stream of particles.

The accepted value of e/m (the charge-to-mass ratio or 'specific charge') for electrons is 1.76×10^{11} C/kg. That is, if you were able to collect 1 kg of electrons, their total charge would be 1.76×10^{11} C. (Of course, you could not collect so many electrons together—they would all repel each other with very large forces. One kilogram of electrons represents a lot of electrons. If every family in Britain had an electric cooker, it would need about 1 kg of electrons from a 240 V supply to flow through their ovens to cook all their Christmas dinners.)

Something quite remarkable emerged from Thomson's measurements. When hydrogen was in the tube, he was able to measure the charge-to-mass ratio for hydrogen ions (emerging through a hole in the cathode). This ratio was about 2000 times less than it was for electrons. This meant that Thomson had discovered a particle which either had about 2000 times the charge of a hydrogen ion, which was considered to be very unlikely, or which had 2000 times less mass (or some combination of the two). The idea that there might be a particle lighter than a hydrogen atom caused considerable stir among some established scientists at the time, who firmly believed in Dalton's atomic theory which said, among other things, that hydrogen atoms were the lightest atoms and they could not be split up. It was a long time before Thomson's ideas were generally accepted.

Thomson also discovered that these particles, which were soon to be called 'electrons', were exactly the same *no matter which gas was in his discharge tube*. In other words, they seemed to be a universal constituent of all matter and also suggested, for the first time, that an atom might have some form of internal structure—a totally novel idea at the time.

The electron was the first of many subatomic particles to be found. What do you think 'subatomic' means?

21.4 The electron gun

Today we are able to make streams of electrons much more easily than Thomson was able to make his cathode rays, using a device called an **electron gun**. Fig. 21.3(a) shows the idea behind the electron gun. There are two metal plates, the **anode** and the **cathode**. The cathode can be heated by a small heater, which is supplied with current from a low-voltage supply (6.3 V is the usual supply voltage). The anode has a hole in it. The whole arrangement is in a vacuum.

Fig. 21.3(b) shows a small variation in the construction. The cathode and heater are combined together, and a small electric current is passed through the cathode itself to make it hot. This is called a **directly-heated cathode**—as opposed to the **indirectly-heated cathode** shown in Fig. 21.3(a).

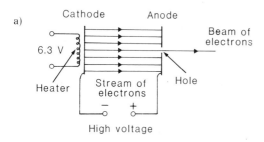

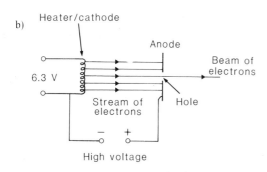

Fig. 21.3   Two kinds of electron gun

When the cathode is heated, electrons in the cathode obtain sufficient energy to escape from the cathode. It is rather like evaporation, except that it is electrons, not atoms, which are evaporating. If the electron gun is connected in a circuit so that the cathode is negative and the anode is positive, the anode will attract the electrons that have 'evaporated' from the cathode, so the electrons will accelerate across to the anode. Most of them will hit

the anode, and so carry on round the circuit back to the power supply. Since electrons have a very small mass, the acceleration is large and they are travelling very fast when they reach the anode—often faster than 10^6 m/s.

When they reach the anode, some electrons will shoot straight through the hole and keep travelling at a high speed until they hit something. Oscilloscopes and televisions contain electron guns. So does the object illustrated in Figs 21.4 and 21.5. This is called a **Maltese cross tube**. When the heater is switched on, it glows white hot, sending out enough light for a shadow of the Maltese cross to be visible on the white screen at the end of the tube, as in the photograph. Round the cathode is a cloud of electrons that have been 'evaporated' off. As soon as a potential difference is applied between the cathode and the anode the electrons are accelerated towards the anode. Some go through a hole in the anode and keep going until they hit either the Maltese cross or the screen. The screen is coated with a special fluorescent substance which glows green when electrons with sufficient energy hit it. The potential difference between the anode and cathode must be at least 1 kV for the electrons to be sufficiently energetic.

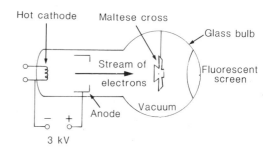

Fig. 21.4   The construction of a Maltese cross tube

Fig. 21.5   A Maltese cross tube

Fig. 21.6   The heater in a Maltese cross tube glows white hot, casting a shadow of the cross on the screen

Fig. 21.8   An electron beam in a deflection tube

Fig. 21.7 shows the result of applying a p.d. between the anode and cathode.

But could this result not be due to a green light which is switched on by the high voltage supply? Try moving a magnet round near the tube—what happens? Could there be a beam of green light in the tube?

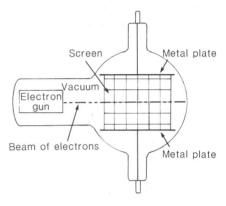

Fig. 21.9   The construction of the deflection tube

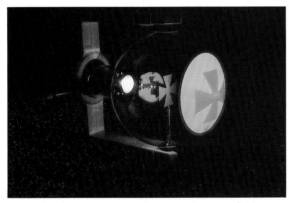

Fig. 21.7   The result of applying a p.d. between the cathode and the anode of the Maltese cross tube

You can see a pair of plates above and below the screen. If the top plate is connected to the positive side of a power supply and the bottom plate to the negative side, thus creating an electric field between the plates, the beam follows the path shown in Fig. 21.10. Explain why.

21.5   Deflecting electrons

The last experiment with a Maltese cross tube and a magnet suggests it is quite easy to push electrons about and deflect them from their path. To investigate this possibility, it would be useful to be able to see the path of the electrons in the tube. There are at least two ways of achieving this. Figs 21.8 and 21.9 show a tube which uses one way. A fluorescent screen is placed diagonally across the tube so that when the beam (which is wide and flat in this tube) strikes the screen it is shown up as a blue line as in the photograph.

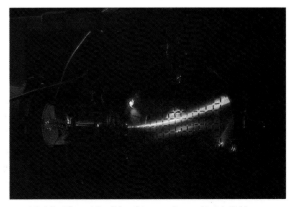

Fig. 21.10   The path of the electrons in an electric field

243

Look carefully at the *shape* of the path. If you have done the work in Section 4.8 you will know this shape is a parabola. If necessary, look back at Section 4.8 to remind yourself of the ideas in that section, then explain why the electron beam does not follow any of the paths illustrated in Fig. 21.11.

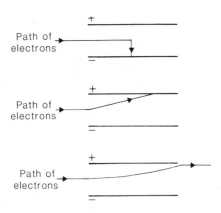

Fig. 21.11   An electron beam cannot follow any of these paths in a uniform electric field

What will happen to the electron beam if the supply to the two sides creating the electric field is reversed?

What is the *maximum* amount by which the electron beam can be deflected by the electric field?

Now try deflecting the electron beam in this tube with a *magnet*. Where do you have to hold the magnet to deflect the beam *down* the screen?

You will probably find it a little difficult to work out what is happening to the beam when you use a magnet to deflect it. This is partly because the strength of the magnetic field is not the same all over the tube, being strongest near the magnet, and partly because you cannot see the path of the beam unless it hits the screen.

To overcome the second problem you can use a tube like the one in Fig. 21.12; this is called a **fine beam tube**. As well as a screen, it contains helium gas at low pressure. When electrons strike the helium atoms, the atoms glow green-blue. There are *two* electron guns in this tube, one pointing vertically up and the other pointing in a horizontal direction as shown in Fig. 21.13. You can see the hole in the conical anode; the cathode and heater are just visible through the bottom of the other anode. A switch on the end of the tube enables you to choose the electron gun you wish to use.

Assemble this tube in its holder and connect it to a suitable power supply. The heater will probably require about 6 V, while the p.d. between anode

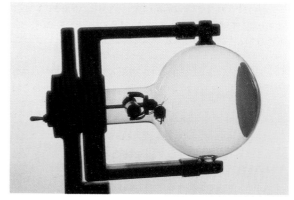

Fig. 21.12   A fine beam tube

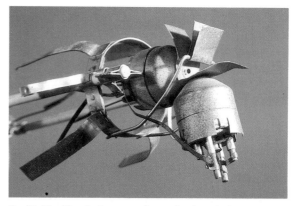

Fig. 21.13   The electron guns from a fine beam tube

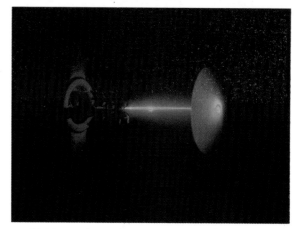

Fig. 21.14   An electron beam crossing the fine beam tube

and cathode will probably need to be about 150 V, but take care to follow the manufacturer's instructions. The green-blue line is very faint, so you will need to black out the laboratory in order to see it.

The line is visible in Fig. 21.14. In this case the horizontal gun was used.

Switch on the *vertical* gun of your tube and bring a magnet near to the gun. Which way does the electron beam move? Turn the magnet round. Which way does the beam move now?

Using *two* magnets can provide quite a lot of fun!

To obtain an *even* magnetic field over the whole tube, it is best to use a special pair of electromagnets which fit round the tube as illustrated in Fig. 21.15. They are called **Helmholtz coils** and are specially designed to give an even magnetic field between the coils.

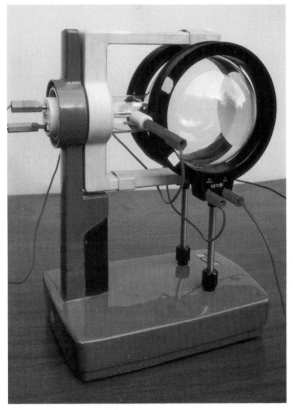

Fig. 21.15 When an electric current is passed through these coils, there will be a magnetic field inside the fine beam tube

The effect of an even magnetic field on the beam is clearly shown in the next photograph—the beam moves in a circle.

Fig. 21.17 The path of the electron beam in a magnetic field

If you have studied the chapter on circular motion, you will know that for something to move in a circle there must be a centripetal force acting on it at *right angles* to the direction of motion. In this case, the force on the electron beam must be at right angles to the direction of motion and the direction of the magnetic field, as shown in Fig. 21.18.

Of course, this idea should not be new to you. Exactly the same angles occurred in the catapult effect, where a wire carrying a current in a magnetic field had a force acting on it perpendicular to the wire and the field. (See Section 17.2.) Since an electric current is just a flow of electrons, it should not surprise you to find the same rule turning up here.

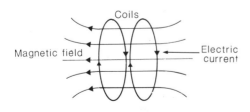

Fig. 21.16 The magnetic field round the coils

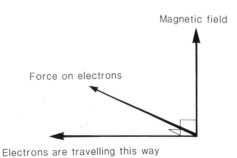

Fig. 21.18 Force, field and movement are all perpendicular to each other

245

21.6 The cathode ray oscilloscope

A **cathode ray oscilloscope** is one of the most useful and versatile pieces of equipment that a physicist uses. It has already been used in this book on a number of occasions and you should now be able to understand how it works.

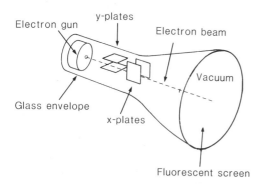

Fig. 21.19 The construction of a cathode ray oscilloscope

Fig. 21.19 shows how a cathode ray oscilloscope is constructed. Electrons are produced by an electron gun at the back of the tube. They then pass between two pairs of plates. Between each pair of plates an electric field can be created. One pair of plates can move the beam up or down, the other pair can move the beam sideways. Thus the beam can be moved so that it can strike the fluorescent screen anywhere. When it strikes the screen, the electron beam produces a spot of light.

Why do you think these plates are called the **x-plates** and the **y-plates**?

The input p.d. to the oscilloscope is connected, via an amplifier, to the y-plates. If, for example, the top y-plate is made positive and the bottom one negative, the beam will bend up (as shown in Fig. 21.20) in the same way as with the deflection tube in the last section. Notice that the path of the beam is only curved while it is between the plates.

Why is the path of the beam straight when it is not between the plates?

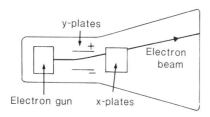

Fig. 21.20 Deflecting the electron beam in a CRO

Draw a diagram like Fig. 21.20 which shows what happens if the bottom plate is positive and the top one negative.

If an alternating p.d. is connected to the oscilloscope, the electron beam will be pulled up, then down, then up again and so on. If this happens fast enough, the spot moves up and down on the screen fast enough to appear as a line.

The x-plates are usually connected to the **time-base circuit** inside the oscilloscope. This is an electronic circuit which produces a voltage which varies in the way shown in Fig. 21.21. It is called a **saw-tooth waveform**.

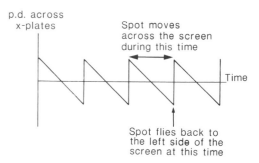

Fig. 21.21 A saw-tooth waveform

At the start of the sequence a fairly large voltage is applied across the x-plates, with the left-hand plate (looking from the front) positive. Which way will the electron beam bend? Where will the electrons hit the screen?

The voltage across the x-plates gradually decreases. What happens to the electron beam? Where will the beam hit the screen when there is no voltage across the x-plates? Which way will the beam bend when the left-hand plate is negative?

If this cycle of events happens very fast, the spot moves too fast to be seen as a spot. What do you see on the oscilloscope screen if there is no p.d. across the y-plates? What do you see if there is an alternating p.d. across the y-plates?

21.7 Geiger and Marsden's experiment

After Thomson discovered the electron, he suggested a model of an atom to try to explain his findings. He suggested that negatively charged electrons were embedded evenly over the main part of the atom, which was positively charged, rather like currants embedded in a currant pudding. This model was called the '**currant pudding**' or '**plum pudding**' model.

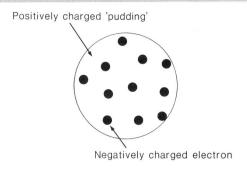

Positively charged 'pudding'

Negatively charged electron

Fig. 21.22   Thomson's 'plum-pudding' model of an atom

The next step forward occurred as a result of an experiment carried out by Geiger and Marsden, working under the direction of Rutherford at Cambridge. They directed alpha particles from a radioactive source at a very thin sheet of gold foil and investigated what happened to the alpha particles. (Gold foil was used because it can be beaten very thin.)

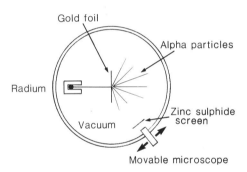

Gold foil

Alpha particles

Radium

Zinc sulphide screen

Vacuum

Movable microscope

Fig. 21.23   Geiger and Marsden's experiment

The vast majority (over 99.9%) of the alpha particles went straight through the gold foil without being deviated from their path at all. Some were deviated by quite a large angle, and, even more surprising, a few 'bounced' back the way they had come. Rutherford was most surprised at this last result. He said it was as if a 15-inch shell had been fired at a piece of tissue paper, and the shell had rebounded from the paper and hit the gunner.

21.8   An atomic model

When Rutherford examined Geiger and Marsden's results, it was clear to him that there must be a lot of *empty space* inside an atom.

What do you think made him come to that conclusion?

Rutherford then reasoned that the alpha particles which were deviated by a large angle, and especially those which turned round, must have had a very large force acting on them. This force could have come from a very concentrated positive charge. The currant pudding model was no use here; the charge was too spread out over all of the 'pudding'. So Rutherford suggested a very small, positively charged nucleus, with a cloud of electrons round the nucleus but a long way from it (Fig. 21.24).

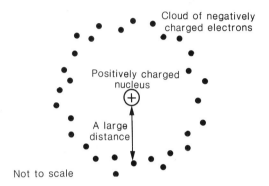

Cloud of negatively charged electrons

Positively charged nucleus

A large distance

Not to scale

Fig. 21.24   Rutherford's atomic model

From the way that the alpha particles were scattered in Geiger and Marsden's experiment, Rutherford was able to calculate the amount of positive charge on the nucleus. From his results, he suggested that the nucleus was made up of many small, positively charged particles which were later called **protons**. The number of protons in an atom depended on which atom it was, and was equal to the number of electrons in the cloud round the nucleus. Since the charge on a proton was equal and opposite to that on an electron, the whole atom was thus electrically neutral.

Unfortunately, this suggestion as it stood did not fit in with the known masses of the atoms. For example, Rutherford suggested that a hydrogen atom had one proton in its nucleus, and one electron. The next more massive atom is helium.

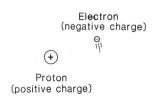

Electron (negative charge)

Proton (positive charge)

Fig. 21.25   Rutherford's model of a hydrogen atom

This was known to have two electrons and therefore, according to Rutherford, must have two protons in its nucleus, and so be twice as massive as hydrogen. But helium atoms were known to be four times as massive as hydrogen atoms. The same problem arose with all other atoms—they were all about twice as massive as Rutherford's simple idea suggested.

One suggestion which was made to overcome this difficulty was that the nucleus should contain both protons and electrons. A helium nucleus, for example, would contain four protons and two electrons. It was later realized that, for fairly advanced reasons (involving 'wave mechanics') that it was unlikely that an electron could be found in a nucleus, and the idea that there was a second particle in the nucleus along with the proton gradually gained favour. This particle had the same mass as a proton but no charge. It was called a **neutron**. A neutron turned out to be very difficult to detect, since it does not do very much. It was not until 1932 that Chadwick obtained experimental evidence for a neutron.

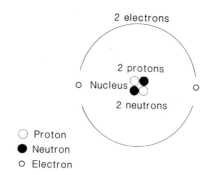

Fig. 21.26　The structure of a helium atom

Physicists eventually concluded, therefore, that all atoms are built up of just three fundamental particles, which are listed in the table below. Protons and neutrons are found in a very compact nucleus, and electrons exist in a cloud round this nucleus, but a long way from it. Most of the atom is empty space.

Particle:	proton	neutron	electron
Mass (compared to proton)	1	1	1/1800
Charge (in units of electronic charge)	+1	0	−1

It is very difficult to imagine how much empty space there is in an atom, and it is impossible to draw an atom to scale on a page of this book. If the full stop at the end of this sentence represents the nucleus of an atom, then, on the same scale, the electron cloud would be at least a metre away. If we wanted to draw a scale diagram which fits this page, the dot to represent the nucleus in the middle would be too small for the printer to be able to print it, and would probably not be visible even if the printer could manage.

21.9　Two definitions

You must understand what the following mean before you can go any further with this chapter.

Proton number

The **proton number** is the number of protons in the nucleus of an atom. (There are the same number of electrons as protons in a neutral atom.)

Nucleon number

Protons and neutrons are collectively called **nucleons**. The **nucleon number** is the total number of particles (that is, protons+neutrons) in the nucleus of an atom.

For example, if the nucleus of an atom of lithium contains three protons and four neutrons, then the proton number is 3 and the nucleon number is 7.

(You may also come across the terms 'atomic number', which means proton number, and 'mass number', which means nucleon number.)

The proton number and nucleon number of an atom is often written just before the symbol for the atom. For example, the symbol

$$^{238}_{92}U$$

refers to an atom of uranium with a nucleon number of 238 and a proton number of 92. The nucleon number should always be written at the top, and the proton number underneath it.

How many neutrons are there in this atom of uranium?

21.10　Isotopes

When Thomson was measuring the charge-to-mass ratio for electrons and also for positive ions, he

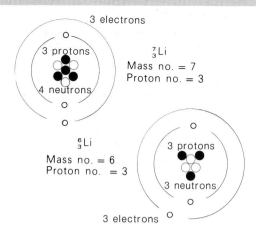

Fig. 21.27   Isotopes of lithium

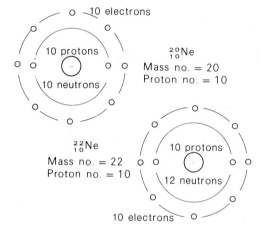

Fig. 21.28   Isotopes of neon

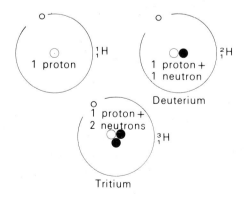

Fig. 21.29   Isotopes of hydrogen

discovered that atoms of the *same* element could have *different* nucleon numbers. For example, he found that neon atoms could have a nucleon number of either 20 or 22.

This was eventually explained as follows. All atoms of a particular element have the same number of *protons* in their nuclei, but not necessarily the same number of neutrons. The number of neutrons has no effect on the chemical properties of an element. For example, all lithium atoms have 3 protons in the nucleus, with 3 electrons surrounding the nucleus. The proton number is 3. Most lithium atoms have 4 neutrons in the nucleus, so the nucleon number is 7, but some lithium atoms have only 3 neutrons (giving a nucleon number of 6), as illustrated in Fig. 21.27. These nuclei are represented by the symbols

$$^{7}_{3}Li \quad \text{and} \quad ^{6}_{3}Li$$

Each form of an element is called a **nuclide**; it is normally referred to by its mass number (e.g. lithium 6, lithium 7). Most elements are found naturally as a mixture of nuclides. For example, neon is a mixture of

$$^{20}_{10}Ne \quad \text{and} \quad ^{22}_{10}Ne$$

Nuclides of the same element with different nucleon numbers (because they have different numbers of neutrons in the nucleus) are called **isotopes**.

Hydrogen has three isotopes (see Fig. 21.29). They are given special names—**hydrogen, deuterium** and **tritium**. Natural hydrogen contains 99.99% 'ordinary' hydrogen atoms and 0.01% deuterium atoms. Tritium is unstable and decays with a half-life of 12.26 years (giving out a low energy beta particle), so there is unlikely to be any in a naturally occurring sample of hydrogen. Tritium can be made artificially in nuclear reactors as the result of a nuclear reaction.

21.11   Explaining radioactivity

Now that you know about the structure of the nucleus of atoms, you should be able to understand what happens when a radioactive atom decays. Radioactivity is entirely to do with the *nucleus* of an atom. Radioactive atoms have unstable nuclei which tend to break down, giving out either an alpha particle or a beta particle, together with some gamma radiation.

Alpha decay

As you know, an alpha particle is the same as a helium nucleus, consisting of two protons and two neutrons. When an atom decays by giving out an alpha particle, the nucleus therefore loses two protons and two neutrons (Fig. 21.30). What is left behind is an entirely new nucleus of a *different element*; this **daughter element** has a proton number *two less* than that of its parent (Why?) and a nucleon number *four less* than that of its parent. (Why?)

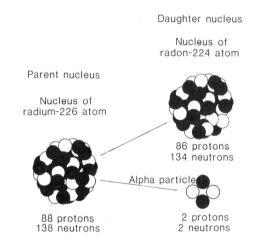

Fig. 21.30   What happens to a nucleus during alpha decay

For example, radium 226 (which has a proton number of 88) decays into radon 222 (which has a proton number of 86). We can write this process in the form of an equation, like this:

$$^{226}_{88}\text{Ra} \longrightarrow ^{222}_{86}\text{Rn} + ^{4}_{2}\text{He}$$

Notice the way in which this equation balances. The total number of particles (i.e. the nucleon number) on the left of the equation is 226, and on the right the total number of particles is the same, $222 + 4 = 226$. The total number of protons on the left is 88 (This is also the total number of units of positive charge—why?) and the total number of protons on the right is $86 + 2 = 88$, i.e. the same.

The radon produced is itself radioactive and decays with the emission of another alpha particle to form an isotope of polonium (Po).

What is the nucleon number of this isotope of polonium? What is the proton number of polonium? Write an equation for the decay of radon into polonium.

Beta decay

How can an electron (i.e. a beta particle) come from a nucleus if the nucleus contains just protons and neutrons? The answer is that a neutron *changes* into a proton and an electron. The proton remains in the nucleus while the electron leaves the nucleus at a high speed.

The total number of particles in the nucleus remains exactly the same during beta decay, so the nucleon number does not change. However, the proton number *increases* by one, so, as with alpha decay, the nucleus of a different element is formed.

For example, the iodine being used in Fig. 21.1 at the start of this chapter contains the radioactive isotope iodine-131. This decays into xenon-131 with the emission of a beta particle.

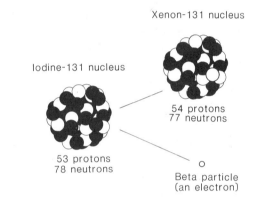

Fig. 21.31   Beta decay

The equation for this nuclear reaction is:

$$^{131}_{53}\text{I} \longrightarrow ^{131}_{54}\text{Xe} + ^{\ 0}_{-1}\beta$$

Notice how we write the symbol for a beta particle in this equation. It can also be written as $^{\ 0}_{-1}\text{e}$. What do you think the 'e' stands for? What does the 0 represent? What does the -1 mean? Why *minus* 1?

Write a similar equation for the beta decay of the radioactive isotope of carbon, carbon-14 (proton number 6) into an isotope of nitrogen.

Gamma emission

Often, when a nucleus emits an alpha or a beta particle, the protons and neutrons left in the nucleus have more energy than normal for that kind of nucleus. The nucleus is said to be in an **excited state**. The protons and neutrons rearrange themselves to make the nucleus more stable, and the extra energy is emitted as gamma radiation.

Gamma emission causes no change in the nucleon number or the proton number.

21.12   Radioactive series

In Section 20.13 you learned that each radioactive substance has its own characteristic half-life. The age of this planet is about 5×10^9 years. How can it be that quantities of nuclides with short half-lives can still be found? Surely any such nuclides should have decayed to negligible quantities. Even the half-life of radium (1622 years) is short compared to the age of this planet. How is it that there is any radium left?

The answer is that many radioactive substances are the *daughters* of longer-lived nuclides. The vast majority of radioactive substances that are found naturally belong to one of three **decay series**. One of these is illustrated in Fig. 21.32. A nucleus of uranium-238 decays successively through all the stages shown until it eventually becomes a nucleus of lead-206, which is stable. Since uranium-238 has a half-life comparable to the age of this planet, there is still about half the original quantity that there was when the planet was formed, so all members of the series are still to be found.

The two other series which are known start with uranium-235 and thorium-232. A fourth series has been predicted, starting with neptunium (proton number 94). Neptunium can be made artificially in a nuclear reactor; it has a half-life of about 2 million years.

Try to think why no natural neptunium is likely to be discovered.

21.13   Nuclear reactions

While Rutherford was working at Cambridge in 1919, he was investigating the effect of bombarding various gases with alpha particles. He discovered that when nitrogen was bombarded with alpha particles, some *protons* were produced, and some of the nitrogen nuclei changed into nuclei of oxygen-17. Rutherford suggested that the equation to represent what was happening is:

$$^{14}_{7}N + ^{4}_{2}He \longrightarrow ^{1}_{1}H + ^{17}_{8}O$$

Later, cloud chamber photographs of this event confirmed Rutherford's suggestion. This was the first artificial nuclear reaction.

The discovery of the neutron by Chadwick in 1932 gave physicists a useful new particle. Since it is uncharged, it is not repelled from the positively charged nucleus of an atom as it approaches (unlike the alpha particles that Rutherford used), so it is easier to make it approach very close to, or even go 'inside', a nucleus.

However, being uncharged, a neutron cannot be accelerated, unlike an alpha particle or a proton. How can you make either of these particles accelerate?

In 1939, two German scientists, following up some work done by Enrico Fermi in Italy, found that if uranium is bombarded with neutrons, the uranium nucleus splits into two smaller nuclei, often, but not always, nuclei of barium and krypton. This process is known as **nuclear fission**.

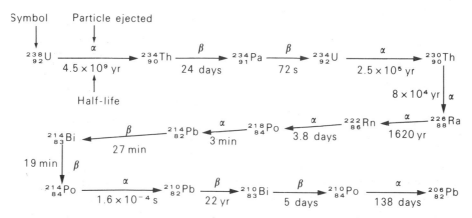

Fig. 21.32   The uranium-238 decay series

The equation for the fission of a uranium-235 nucleus can be written:

$$^{235}_{92}U + ^{1}_{0}n \longrightarrow ^{141}_{56}Ba + ^{92}_{36}Kr + 3^{1}_{0}n$$

A lot of energy is given out in this process, most of which appears as kinetic energy of the barium and krypton nuclei. If all the nuclei in 1 gram of uranium-235 underwent fission, the energy released would be about the same as that released on burning $2\frac{1}{2}$ tonnes of coal.

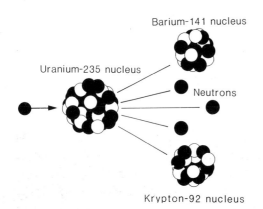

Barium-141 nucleus

Uranium-235 nucleus

Neutrons

Krypton-92 nucleus

Fig. 21.33   Splitting a nucleus of uranium-235

As illustrated in Fig. 21.33, when a uranium-235 nucleus splits it gives out some neutrons (usually 3) as well as the nuclei of atoms like barium and krypton. If the total volume of uranium is fairly small, most of these neutrons will pass through the uranium atoms and escape. However, with a large volume of uranium, there is a large chance of more than one of the neutrons being 'captured' by other nuclei, causing them to split, so a chain reaction will occur, as illustrated in Fig. 21.34. This can rapidly result in an explosion. The mass of uranium

required to *just* sustain a chain reaction, where, on average, one out of three neutrons is captured by another uranium nucleus (Fig. 21.35), is called the **critical mass**. If a chain reaction is just sustained, the average number of nuclei fissioning at any one time remains constant.

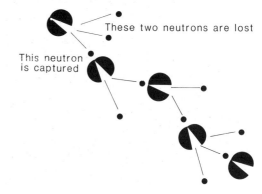

These two neutrons are lost

This neutron
is captured

Fig. 21.35   A critical system

Unfortunately, the first use that was made of the energy that can be released from a splitting uranium-235 nucleus was in the atomic bomb that was dropped on Hiroshima in 1945.

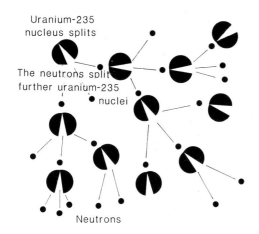

Uranium-235
nucleus splits

The neutrons split
further uranium-235
nuclei

Neutrons

Fig. 21.34   A chain reaction

Fig. 21.36   The atomic explosion at Nagasaki in 1945

Fig. 21.37 The effect of an atomic bomb on Nagasaki

21.14 Nuclear power

The power station illustrated in Fig. 21.38 is using energy from splitting uranium-235 nuclei to generate electricity. In many ways it is like a coal-fired power station. In both kinds of power stations energy stored in the fuel is used to heat water which produces steam which drives turbines which turn generators which generate electricity. The energy is produced by burning coal in coal-fired power stations; fission provides the energy in a nuclear power station.

Fig. 21.38 Trawsfynydd nuclear power station

Natural uranium consists of mostly uranium-238, only 0.7% being uranium-235. Uranium-238 does undergo fission, but the neutrons must be moving very fast for this to occur—faster than those which come from a uranium-235 nucleus when it undergoes fission. Uranium-235 will undergo fission with neutrons of any energy, but the process is most efficient with low energy neutrons. Natural uranium had to be 'enriched' with a much higher concentration of uranium-235 before it could be used in a bomb. The bomb contained two subcritical masses of enriched uranium. When they were brought together, the uranium became super-critical and exploded.

Fig. 21.39 shows the main parts of a nuclear power station. The fuel elements contain natural uranium. Neutrons produced by fission are slowed down by a **moderator** (usually graphite) so that they have a good chance of causing the fission of further uranium-235 nuclei, so that a chain reaction keeps going.

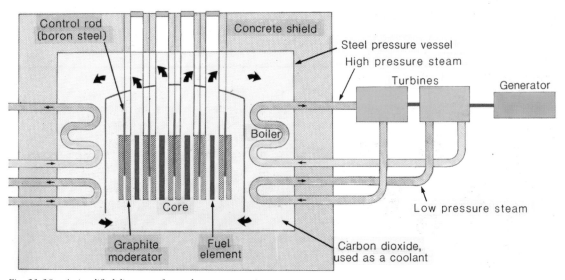

Fig. 21.39 A simplified diagram of a nuclear power station

The number of neutrons that are available for causing fission, and therefore the temperature of the reactor, is controlled by the **control rods**. These are usually made of boron, which is good at *absorbing* neutrons. If they are lowered into the reactor core, more neutrons are absorbed and less fission can take place. They can be dropped right down into the core to stop the chain reaction quickly in an emergency.

The energy from the reactor core is removed by a coolant. This might be carbon dioxide (as in the Magnox and Advance Gas Cooled reactors at present producing electricity in Britain) or water at high pressure (as in the pressurized water reactor, which is used in the USA and which, at the time of writing this book, the United Kingdom Atomic Energy Authority [UKAEA] wishes to build in Britain).

The hot gas, or water, boils water in a boiler, and the steam produced drives a generator in the usual way.

Why do you think the core is surrounded by a steel pressure vessel and a thick concrete shield?

Only the uranium-235 in natural uranium undergoes fission in a nuclear reactor. Eventually so much of the uranium-235 has been turned into other products that the fuel element has to be replaced (Fig. 21.40). However, the uranium-238, although it does not undergo fission (since the neutrons are moving too slowly) can *capture* a neutron, turning into uranium-239, which decays to neptunium-239 and then to plutonium-239.

$$^{1}_{0}n + ^{238}_{92}U \longrightarrow ^{239}_{92}U \longrightarrow ^{239}_{93}Np \longrightarrow ^{239}_{94}Pu$$

What particles must be given out from the uranium-239 and the neptunium-239 nuclei as they decay?

Plutonium-239 is fissionable. It was the material in the atomic bomb dropped on Nagasaki in 1945. About 1% of the 'spent' fuel from a nuclear reactor consists of plutonium. Until recently, the main use of this plutonium was to make atomic bombs, but a new kind of reactor, the **fast-breeder reactor**, uses a mixture of uranium-238 and plutonium. The plutonium is sufficiently concentrated for the fission reaction to be sustained and no moderator is used since fast neutrons are needed. What is particularly useful is that the uranium captures neutrons and changes to plutonium as described above. As the original plutonium is used up, more is created from the uranium-238. Hence the reason for the term 'breeder' reactor—it breeds its own fuel.

Are we likely to use more nuclear power in Britain in the future? At present, nuclear power stations provide over 10% of our electricity. Supplies

Fig. 21.40
View inside the core of a nuclear reactor

Fig. 21.41 Control rod mechanisms for a nuclear reactor

of oil and gas will rapidly decline at the end of this century. There are large resources of coal, but this will be needed for other purposes than burning in power stations (What, for example?). Nuclear power is cheap. The following figures from the CEGB, dated December 1979, show the cost per kW h of electricity generated by three different fuels.

Energy source	Cost per kW h
nuclear (Magnox)	1.02p
coal	1.29p
oil	1.31p

But world resources of uranium are limited. 'Thermal' reactors (Magnox and Advanced Gas

Cooled) use only 0.7% of the fuel (i.e. the uranium-235), whch is wasteful. Breeder reactors would use much of the remaining fuel, converting it to fissionable plutonium as described above. Britain has so far stockpiled about 20 000 tonnes of spent uranium. Used in breeder reactors, this could supply the same energy as about 400 years worth of coal.

What are the disadvantages of nuclear power stations? There are two main points which cause people concern. One is that plutonium is a basic requirement for nuclear weapons. The other is the fact that the fission products (for example, barium and krypton) are highly radioactive, and have long half-lives. They have to be stored safely for thousands of years, and be kept cool, since the energy released during radioactive decay causes an increase in temperature. How can we be sure these waste products remain safe? Are we right to pass on to future generations a storage problem of this nature? Are the proposals that have been made to dump the waste material in the sea (in thick concrete containers) or under the ground sensible and safe?

Are the hazards involved in nuclear power significantly greater than those involved with a coal-fired power station? A typical 1000 MW coal-fired power station will, for example, produce over 100 000 tonnes of sulphur dioxide in a year. This causes bronchial and chest diseases and also 'acid rain'. What is 'acid rain' and why is it a hazard? The same power station will produce over 27 000 tonnes of oxides of nitrogen in a year; oxides of nitrogen can cause cancer.

Eventually people will have to decide whether nuclear power is worthwhile. One factor is certain to influence the decision. Most people will not take kindly to the idea that their standard of living will go down because it is not possible to generate enough electricity.

21.15 Nuclear fusion

There is a second way of obtaining energy from nuclei. If two light nuclei, such as hydrogen nuclei, are pushed together hard enough, they will fuse together to form a new more massive, nucleus, giving out some energy in the process. Fusion of hydrogen nuclei is believed to be the source of the Sun's energy.

The fusion of deuterium ('heavy hydrogen') nuclei can be represented by this nuclear equation:

$$\,_1^2\mathrm{H} + \,_1^2\mathrm{H} \longrightarrow \,_2^3\mathrm{He} + \,_0^1\mathrm{n}$$

A lot of research has taken place over many years to try to carry out this particular reaction under controlled conditions. There are many problems yet to be overcome, but if physicists are ever successful, then many of the world's energy problems could be solved, for there is an almost unlimited supply of fuel in the sea.

Fig. 21.42   Tokamak nuclear fusion experiment, Culham Laboratory, Oxfordshire

21.16   The end of classical physics

All the physics in this book has been what is generally known as *classical* physics. You have learned, for example, about Newtonian mechanics, and you have seen examples of energy carried by waves. Since Rutherford's time, physicists have discovered many other facts about atoms, and they have tried to explain these facts in terms of the ideas of classical physics, such as Newtonian mechanics and waves. They failed. They could not satisfactorily explain all they discovered about atoms using the ideas you have learned about in this book. New ideas and new theories were needed.

The beginning of the twentieth century saw the introduction of new, 'modern' ideas such as quantum mechanics and relativity. This does not mean that the old, classical, ideas were wrong; they are just inadequate to explain all of physics. It is at this junction between classical and modern physics that this book must stop. Do not think you have wasted your time learning about physics in this book! It is still perfectly adequate for describing the everyday behaviour of large-scale objects.

SUMMARY

Now that you have finished studying this chapter on inside the atom, there are a number of things you should know and be able to do.

1 You should:
   a) be able to explain how an electron gun produces a stream of electrons;
   b) know that electrons can be deflected in a magnetic field, and be able to predict the direction in which they will be deflected;
   c) know that electrons can be deflected in an electric field, and be able to predict the direction in which they will be deflected;
   d) be able to describe how a cathode ray oscilloscope works;
   e) know the significance of Thomson's discovery of the electron;
   f) understand the principle of the Geiger and Marsden experiment;
   g) understand how the results of the Geiger and Marsden experiment show there is a lot of space in an atom;
   h) know that an atom consists of a concentrated nucleus surrounded by clouds or shells of electrons;
   i) know that a nucleus contains both protons and neutrons;
   j) know the meaning of proton number and nucleon number;

   k) be able to explain what is happening to the nucleus of an atom during alpha, beta and gamma emission;
   l) understand what happens during the fission of a uranium nucleus;
   m) understand the principle of a nuclear power station;
   n) understand what nuclear fusion is.

2 You should know what each of the following is, or what each does:

cathode	anode
Maltese cross tube	fine beam tube
Helmholtz coils	saw-tooth waveform
deuterium	tritium
excited state	radioactive series
daughter product	critical mass
moderator	control rods
acid rain	

3 These are some of the other words that have been used in this chapter. You should know what each word means:

evaporate	fluorescent
conical	universal
constituent	fundamental
comparable	stockpile
bronchial	

FURTHER QUESTIONS

1 The isotope of lead, $^{214}_{82}Pb$, is radioactive. When it decays, it gives out a β-particle and forms an isotope of bismuth.
   a) What does the number 214 represent?
   b) What does the number 82 represent?
   c) If the isotope of bismuth can be written $^{x}_{y}Bi$, what are the correct values of x and y?
   d) Write a nuclear equation for the decay of lead.

2 Part of the uranium-238 decay series is shown in Fig. 21.43.
   a) Explain the meaning of the symbol $^{238}_{92}U$.
   b) What is the symbol for an alpha particle?
   c) What particle is given out at each step of the part of the decay series in Fig. 21.43?
   d) Write down all the pairs of isotopes which are present in the part of the series in Fig. 21.43.

Symbol	Half life
$^{238}_{92}U$	
↓	4.5×10^9 yr
$^{234}_{90}Th$	
↓	24 days
$^{234}_{91}Pa$	
↓	72 s
$^{234}_{92}U$	
↓	2.5×10^5 yr
$^{230}_{90}Th$	

Fig. 21.43

e) If the end-product of the complete series is lead-206, how many alpha particles are emitted in the complete decay series, from uranium-238 to lead-206?

3 The specific charge of an electron is 1.76×10^{11} C/kg. The specific charge of a proton is 9.6×10^8 C/kg.
   a) What do you think the phrase 'specific charge' means?
   b) Explain why it is reasonable to assume that an electron and a proton carry equal but opposite charges.
   c) If the charge on an electron is 1.9×10^{-19} C, what is the mass of a proton?

4 Fig. 21.44 shows a hollow glass sphere containing a gas at very low pressure. An electron gun, consisting of a filament F and an anode A, is connected as shown. Electrons from the gun follow the path AB, which can be seen as a blue line inside the sphere.

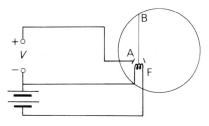

Fig. 21.44

a) What is the purpose of the filament?
b) What is the purpose of the anode?
c) If the potential difference between F and A is V, how much energy does this potential difference give to an electron having a charge e?
d) If the kinetic energy of an electron of mass m, leaving the gun with a speed v, is $\frac{1}{2}mv^2$, find an expression for v in terms of e, V and m. Explain your reasoning.

A second circuit provides a magnetic field and the electrons are then seen to follow a circular path, as shown in Fig. 21.45.

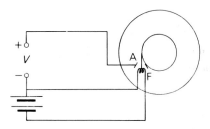

Fig. 21.45

e) Draw a circuit which can be used to provide a magnetic field which can be varied in strength.
f) What does the fact that the electron travels in a circular path tell you about:
   i) the direction of the force acting on each electron;
   ii) the magnetic field?
g) How would the radius of the circular path change if:
   i) the magnetic field were increased;
   ii) the voltage V were increased?

After watching such an experiment, members of the class were heard to say:
   'What happens to all the electrons in the end?'
   'I saw this experiment at my last school and the electrons were a different colour.'
h) Write comments on each of the above remarks.
(Nuffield O-Level Physics, 1979)

5 A manufacturer makes an electron 'deflection tube' which consists of an evacuated glass bulb with some electrodes in it (Fig. 21.46). He states that the tube works best if a 6 V supply is connected between A and B and a 3 kV supply between A and C. The electrode at C has a small hole in it, and just beyond the hole there are two parallel metal plates with a connection to each of them.

a) What is the purpose of having a potential difference between A and B?

b) Which terminal of the 3 kV supply should be connected to C?

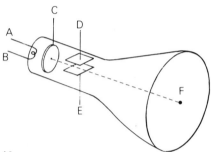

Fig. 21.46

c) Explain the purpose of the high voltage.

d) Why is the tube 'evacuated'?

When the manufacturer's instructions are followed, a spot of green light appears at F on a special coating on the inside of this deflection tube.

e) When a p.d. of 100 V is applied between D and E (D positive) the spot moves away from F. Draw a diagram to show the direction in which the spot moves.

f) Explain why the spot moves in e).

g) How could you make the spot move further from F?

h) If the spot is not very bright, what *two* separate changes might you make to increase the brightness?

i) An alternating voltage of frequency 50 Hz is applied across the plates D and E. Describe what you now see on the screen.

j) The tube shown in Fig. 21.46 could *not* be used to display an alternating voltage as a wavy line, as an oscilloscope does. What could be done to achieve this result?

(Nuffield O-Level Physics, 1977)

APPENDIX– QUANTITIES AND UNITS

The following is a list of the various physical quantities that have been used in this book, together with the equation (where appropriate) which is used to calculate that quantity, and the units in which that quantity is measured.

Quantity	Calculated from	Measured in	Symbol for unit
distance	—	metre	m
time	—	second	s
mass	—	kilogram	kg
density	$\dfrac{\text{mass}}{\text{volume}}$	kilogram per cubic metre	kg/m^3
velocity	$\dfrac{\text{distance}}{\text{time}}$	metres per second	m/s
acceleration	$\dfrac{\text{change of velocity}}{\text{time}}$	metres per second squared	m/s^2
force	mass × acceleration	newton	N
pressure	$\dfrac{\text{force}}{\text{area}}$	pascal	Pa
impulse	force × time	newton second	N s
momentum	mass × velocity	kilogram metres per second	kg m/s
energy transferred	force × distance	joule	J
power	$\dfrac{\text{energy}}{\text{time}}$	watt	W
current	—	ampere	A
charge	current × time	coulomb	C
potential difference	$\dfrac{\text{energy}}{\text{charge}}$	volt	V
resistance	$\dfrac{\text{potential difference}}{\text{current}}$	ohm	Ω

The following are the standard prefixes used to denote multiples or divisions of a unit:

tera	T	10^{12}
giga	G	10^{9}
mega	M	10^{6}
kilo	k	10^{3}
milli	m	10^{-3}
micro	μ	10^{-6}
nano	n	10^{-9}
pico	p	10^{-12}

For example:

$1000\,\text{m} = 10^{3}\,\text{m} = 1 \text{ kilometre} = 1\,\text{km}$

$1\,000\,000\,\text{W} = 10^{6}\,\text{W} = 1 \text{ megawatt} = 1\,\text{MW}$

$\frac{1}{1000}\,\text{A} = 10^{-3}\,\text{A} = 1 \text{ milliamp} = 1\,\text{mA}$

The following are some of the formulae and relationships that have been used in this book.

Bodies in motion

$v = u + at$

$s = ut + \frac{1}{2}at^2$

$v^2 = u^2 + 2as$

$F = ma$

$Ft = mv - mu$

$F = \dfrac{mv^2}{r}$

where:
u = initial velocity
v = final velocity
s = distance travelled in a straight line (displacement)
t = time
a = acceleration
F = force
m = mass
r = radius of circle

Energy

gravitational potential energy $= mgh$

elastic potential energy $= \frac{1}{2}Tx$

kinetic energy $= \frac{1}{2}mv^2$

energy $=$ mass $\times$ specific heating energy $\times$ temperature change

electrical energy $=$ potential difference $\times$ charge

where:
m = mass
g = gravitational field strength
h = vertical height
T = tension
x = extension
v = velocity

Electricity

power = potential difference $\times$ current

potential difference = current $\times$ resistance

Waves

velocity = frequency $\times$ wavelength

For Young's double slits:

$$\text{wavelength} = \frac{\text{slit separation} \times \text{slit width}}{\text{distance from slits to screen}}$$

General gas law

$$\frac{P_1 V_1}{T_1} = \frac{P_2 V_2}{T_2}$$

where:
P = pressure
V = volume
T = temperature

Expansion of solids

change in length $=$ expansivity $\times$ length $\times$ temperature change

INDEX